SEX

Margaret Mead's famous study of adolescence and sex in primitive society is a scientific classic that has been widely acclaimed by both educators and the general reader.

Coming of Age in Samoa is a frank and beautiful book, incisive and original, leavened with charm and humor. Dr. Mead describes the basic principles of Samoan life which shape attitudes and behavior from birth to maturity, and vividly portrays the moral and social problems adolescents have to solve and the values that guide them in their solutions.

Among the subjects Dr. Mead discusses are: education, family relationships and rivalry, community life, pre-marital and post-marital sex taboos, homosexuality, the role of the dance, personality and temperament, and religion.

"As Miss Mead's careful scientific work deserves the most earnest tribute, so her method of presenting its results calls for the highest praise. Her book, broad in its canvas and keen in its detail, is sympathetic throughout, warmly human, yet never sentimental, frank with the clean, clear frankness of the scientist, unbiased in its judgment, richly readable in its style. It is a remarkable contribution to our knowledge of humanity." —*New York Times*

THIS IS A REPRINT OF THE ORIGINAL HARDCOVER EDITION PUBLISHED BY WILLIAM MORROW & COMPANY

COMING OF AGE IN
SAMOA

*A Psychological Study of Primitive
Youth for Western Civilisation*

MARGARET MEAD
Foreword by Franz Boas

A MENTOR BOOK
Published by THE NEW AMERICAN LIBRARY

To the Girls of Tau
This Book Is Dedicated

'Ou te avatu
lenei tusitala
ia te 'outou
O Teinetiti ma le Aualuma
o Taū

Contents

Acknowledgments

I am indebted to the generosity of the Board of Fellowships in the Biological Sciences of the National Research Council whose award of a fellowship made this investigation possible. I have to thank my father for the gift of my travelling expenses to and from the Samoan Islands. To Professor Franz Boas I owe the inspiration and the direction of my problem, the training which prepared me to undertake such an investigation, and the criticism of my results.

For a co-operation which greatly facilitated the progress of my work in the Pacific, I am indebted to Dr. Herbert E. Gregory, Director of the B. P. Bishop Museum and to Dr. E. C. S. Handy and Miss Stella Jones of the Bishop Museum.

To the endorsement of my work by Admiral Stitt and the kindness of Commander Owen Mink, U. S. N., I owe the co-operation of the medical authorities in Samoa, whose assistance greatly simplified and expedited my investigation. I have to thank Miss Ellen M. Hodgson, Chief Nurse, the staff nurses, the Samoan nurses, and particularly G. F. Pepe for my first contacts and my instruction in the Samoan language. To the hospitality, generosity, and sympathetic co-operation of Mr. Edward R. Holt, Chief Pharmacist Mate, and Mrs. Holt, I owe the four months' residence in their home which furnished me with an absolutely essential neutral base from which I could study all the individuals in the village and yet remain aloof from native feuds and lines of demarcation.

The success of this investigation depended upon the co-operation and interest of several hundred Samoans. To mention each one individually would be impossible. I owe special thanks to County Chief Ufuti of Vaitogi and to all the members of his household and to the Talking Chief Lolo, who taught me the rudiments of the graceful pattern of social relations which is so characteristic of the Samoans. I must specially thank their excellencies, Tufele, Governor of Manu'a, and County Chiefs Tui Olesega, Misa, Sotoa, Asoao, and Leui, the Chiefs Pomele, Nua, Tialigo, Moa, Maualupe, Asi, and the Talking Chief Lapui and Muao; the Samoan pastors, Solomona and Iakopo; the Samoan teachers, Sua, Napoleon, and Eti; Toaga, the wife of Sotoa, Fa'apua'a, the Taupo of Fitiuta, Fofoa, Laula, Leauala, and Felofiaiana, and the chiefs and people of all the villages of Manu'a and their children. Their kindness, hospitality, and courtesy made my sojourn among them a happy one; their co-operation and interest made it possible for me to pursue my investigation with peace and profit. The fact that no real names are used in the course of the book is to shield the feelings of those who would not enjoy such publicity.

For criticism and assistance in the preparation of this manuscript I am indebted to Dr. R. F. Benedict, Dr. L. S. Cressman, Miss M. E. Eichelberger, and Mrs. M. L. Loeb.
The American Museum of Natural History, March, 1928. M. M.

Foreword

MODERN descriptions of primitive people give us a picture of their culture classified according to the varied aspects of human life. We learn about inventions, household economy, family and political organisation, and religious beliefs and practices. Through a comparative study of these data and through information that tells us of their growth and development, we endeavour to reconstruct, as well as may be, the history of each particular culture. Some anthropologists even hope that the comparative study will reveal some tendencies of development that recur so often that significant generalisations regarding the processes of cultural growth will be discovered.

To the lay reader these studies are interesting on account of the strangeness of the scene, the peculiar attitudes characteristic of foreign cultures that set off in strong light our own achievements and behaviour.

However, a systematic description of human activities gives us very little insight into the mental attitudes of the individual. His thoughts and actions appear merely as expressions of rigidly defined cultural forms. We learn little about his rational thinking, about his friendships and conflicts with his fellowmen. The personal side of the life of the individual is almost eliminated in the systematic presentation of the cultural life of the people. The picture is standardised, like a collection of laws that tell us how we should behave, and not how we behave; like rules set down defining the style of art, but not the way in which the artist elaborates his ideas of beauty; like a list of inventions, and not the way in which the individual overcomes technical difficulties that present themselves.

And yet the way in which the personality reacts to culture is a matter that should concern us deeply and that makes the studies of foreign cultures a fruitful and useful field of research. We are accustomed to consider all those actions that are part and parcel of our own culture, standards which we follow automatically, as common to all mankind. They are deeply ingrained in our behaviour. We are moulded in their forms so that we cannot think but that they must be valid everywhere.

Courtesy, modesty, good manners, conformity to definite

ethical standards are universal, but what constitutes courtesy, modesty, good manners, and ethical standards is not universal It is instructive to know that standards differ in the most unexpected ways. It is still more important to know how the individual reacts to these standards.

In our own civilisation the individual is beset with difficulties which we are likely to ascribe to fundamental human traits. When we speak about the difficulties of childhood and of adolescence, we are thinking of them as unavoidable periods of adjustment through which every one has to pass. The whole psycho-analytic approach is largely based on this supposition.

The anthropologist doubts the correctness of these views, but up to this time hardly any one has taken the pains to identify himself sufficiently with a primitive population to obtain an insight into these problems. We feel, therefore, grateful to Miss Mead for having undertaken to identify herself so completely with Samoan youth that she gives us a lucid and clear picture of the joys and difficulties encountered by the young individual in a culture so entirely different from our own. The results of her painstaking investigation confirm the suspicion long held by anthropologists, that much of what we ascribe to human nature is no more than a reaction to the restraints put upon us by our civilisation.

FRANZ BOAS

Preface to the 1949 Edition

THIS WAS my first book, the record of my first field trip, at the age of twenty-three, almost a quarter of a century ago, when it took not twenty-four hours, but two weeks to reach Samoa. The years between, the seven other peoples which I have studied, the other languages which I have learned, have not served to dim the special quality of that first attempt to see the life of a very different people, both as they saw it and as they could never see it. I had decided to become an anthropologist—in May, 1923—because Franz Boas and Ruth Benedict had presented the tasks of anthropology as more urgent than any other task which lay ready to the choice of a student of human behavior. Anthropology was a little-known and slightly developed science; there were only a handful of field workers in the world, yet the material with which they worked—the cultures of people who by historical accident had developed ways of life which contrasted with, and so revealed our own—was perishing rapidly. Each epidemic, each new discovery of oil or gold, each new spurt of missionary endeavor brought another people into the circle of our western way of thinking, blotting out forever the living record of what their way had once been. It was not, however, until a year later, at the meeting of the British Association of Science in 1924—where I read my first scientific paper—in talking with a group of anthropologists from many places, each of whom spoke with sure detail of "my people," or even of "among several of my people," that field work became a vivid desire rather than a mere necessary part of a chosen career.

From then on, all the exercises through which aspiring young anthropologists must go to be ready to plunge all alone into the language and customs of an alien people, for which they then become finally responsible, seemed like marking time. Yet this eagerness was very thoroughly tempered with apprehension, especially because I had been selected to do a different kind of field work, on a different kind of problem than any that had been worked on hitherto. For my first two months in Samoa, as I learned to speak the language, eat the food, and use and interpret the postures and the gestures of the people, I found myself often saying under my breath, "I can't do it. I can't do it." One day I noticed that

I was no longer saying this in English but in Samoan, and then I knew that I could.

Whether a book is dated or not, as well as whether it will mean something useful, perhaps, but very different, when read by a different generation from those for whom it was written depends on many things. To the extent that the anthropologist records the whole picture of any way of life, that record cannot fade, because it is the way of life itself, the coherent creation of many generations which keeps, unchanging, the vigor and beauty which belongs to it. Once written down, what might otherwise have survived only in a few quaint customs of a people become part of the wider world, can become a precious permanent possession.

Yet however faithful the record, each anthropologist works with the ideas of his particular period. This book was written when it seemed very necessary to stress that much of what we took for granted, as given by human or even racial inheritance, was learned, and depended upon the teaching of one generation by another for its perpetuation. The very word *culture,* now part of the working equipment of a whole group of sciences, was then a technical term in a small specialist field. Here and there throughout this book, those who have been born since it was written will stop and wonder, why this emphasis, why that insistency. If they ponder a little on their very wonder, they will become by this surer inheritors of these paths which we cut with a sense of freshness, which they inherit as well-trodden. Finally, if the morals I drew for our society, especially for our forms of education, remain—as they seem to when I reread them—pertinent to our world today, that is because I had the unusual privilege of having a grandmother and a mother who lived so close to the growing edge of civilization that I was also able to reflect the determination with which they lived in a not yet existent but foreseeable world.

MARGARET MEAD

July 9, 1949

1

Introduction

DURING the last hundred years parents and teachers have ceased to take childhood and adolescence for granted. They have attempted to fit education to the needs of the child, rather than to press the child into an inflexible educational mould. To this new task they have been spurred by two forces, the growth of the science of psychology, and the difficulties and maladjustments of youth. Psychology suggested that much might be gained by a knowledge of the way in which children developed, of the stages through which they passed, of what the adult world might reasonably expect of the baby of two months or the child of two years. And the fulminations of the pulpit, the loudly voiced laments of the conservative social philosopher, the records of juvenile courts and social agencies all suggested that something must be done with the period which science had named adolescence. The spectacle of a younger generation diverging ever more widely from the standards and ideals of the past, cut adrift without the anchorage of respected home standards or group religious values, terrified the cautious reactionary, tempted the radical propagandist to missionary crusades among the defenseless youth, and worried the least thoughtful among us.

In American civilisation, with its many immigrant strains, its dozens of conflicting standards of conduct, its hundreds of religious sects, its shifting economic conditions, this unsettled, disturbed status of youth was more apparent than in the older, more settled civilisation of Europe. American conditions challenged the psychologist, the educator, the social philosopher, to offer acceptable explanations of the growing children's plight. As to-day in post-war Germany, where the younger generation has even more difficult adjustments to make than have our own children, a great mass of theorising about adolescence is flooding the book shops; so the psychologist in America tried to account for the restlessness of youth. The result was works like that of Stanley Hall on "Adolescence," which ascribed to the period through which the children were passing, the causes of their conflict and distress. Adolescence was characterised as the period in which idealism flowered and rebellion against authority waxed strong, a period during which difficulties and conflicts were absolutely inevitable.

The careful child psychologist who relied upon experiment for his conclusions did not subscribe to these theories. He said, "We have no data. We know only a little about the first few months of a child's life. We are only just learning when a baby's eyes will first follow a light. How can we give definite answers to questions of how a developed personality, about which we know nothing, will respond to religion?" But the negative cautions of science are never popular. If the experimentalist would not commit himself, the social philosopher, the preacher and the pedagogue tried the harder to give a short-cut answer. They observed the behaviour of adolescents in our society, noted down the omnipresent and obvious symptoms of unrest, and announced these as characteristics of the period. Mothers were warned that "daughters in their teens" present special problems. This, said the theorists, is a difficult period. The physical changes which are going on in the bodies of your boys and girls have their definite psychological accompaniments. You can no more evade one than you can the other; as your daughter's body changes from the body of a child to the body of a woman, so inevitably will her spirit change, and that stormily. The theorists looked about them again at the adolescents in our civilisation and repeated with great conviction, "Yes, stormily."

Such a view, though unsanctioned by the cautious experimentalist, gained wide currency, influenced our educational policy, paralysed our parental efforts. Just as the mother must brace herself against the baby's crying when it cuts its first tooth, so she must fortify herself and bear with what equanimity she might the unlovely, turbulent manifestations of the "awkward age." If there was nothing to blame the child for, neither was there any programme except endurance which might be urged upon the teacher. The theorist continued to observe the behaviour of American adolescents and each year lent new justification to his hypothesis, as the difficulties of youth were illustrated and documented in the records of schools and juvenile courts.

But meanwhile another way of studying human development had been gaining ground, the approach of the anthropologist, the student of man in all of his most diverse social settings. The anthropologist, as he pondered his growing body of material upon the customs of primitive people, grew to realise the tremendous rôle played in an individual's life by the social environment in which each is born and reared. One by one, aspects of behaviour which we had been accustomed to consider invariable complements of our humanity were found to be merely a result of civilisation, present in the inhabitants of one country, absent in another country, and this without a change of race. He learned that neither race nor common humanity can be held responsible for many of the forms which even such basic human emotions as love and fear and anger take under different social conditions.

So the anthropologist, arguing from his observations of the behaviour of adult human beings in other civilisations, reaches many of the same conclusions which the behaviourist reaches in his work upon human babies who have as yet no civilisation to shape their malleable humanity.

With such an attitude towards human nature the anthropologist listened to the current comment upon adolescence. He heard attitudes which seemed to him dependent upon social environment—such as rebellion against authority, philosophical perplexities, the flowering of idealism, conflict and struggle —ascribed to a period of physical development. And on the basis of his knowledge of the determinism of culture, of the plasticity of human beings, he doubted. Were these difficulties due to being adolescent or to being adolescent in America?

For the biologist who doubts an old hypothesis or wishes to test out a new one, there is the biological laboratory. There, under conditions over which he can exercise the most rigid control, he can vary the light, the air, the food, which his plants or his animals receive, from the moment of birth throughout their lifetime. Keeping all the conditions but one constant, he can make accurate measurement of the effect of the one. This is the ideal method of science, the method of the controlled experiment, through which all hypotheses may be submitted to a strict objective test.

Even the student of infant psychology can only partially reproduce these ideal laboratory conditions. He cannot control the pre-natal environment of the child whom he will later subject to objective measurement. He can, however, control the early environment of the child, the first few days of its existence, and decide what sounds and sights and smells and tastes are to impinge upon it. But for the student of the adolescent there is no such simplicity of working conditions. What we wish to test is no less than the effect of civilisation upon a developing human being at the age of puberty. To test it most rigorously we would have to construct various sorts of different civilisations and subject large numbers of adolescent children to these different environments. We would list the influences the effects of which we wished to study. If we wished to study the influence of the size of the family, we would construct a series of civilisations alike in every respect except in family organisation. Then if we found differences in the behaviour of our adolescents we could say with assurance that size of family had caused the difference, that, for instance, the only child had a more troubled adolescence than the child who was a member of a large family. And so we might proceed through a dozen possible situations—early or late sex knowledge, early or late sex-experience, pressure towards precocious development, discouragement of precocious development, segregation of the sexes or coeducation from infancy, division of labour between the sexes or common tasks for both, pressure to make religious choices young or the lack of such pressure. We would vary one factor, while the others remained quite constant, and analyse which, if any, of the aspects of our civilisation were responsible for the difficulties of our children at adolescence.

Unfortunately, such ideal methods of experiment are denied to us when our materials are humanity and the whole fabric of a social order. The test colony of Herodotus, in which babies were to be isolated and the results recorded, is not a possible approach. Neither is the method of selecting from our own civilisation groups of children who meet one requirement or another. Such a method would be to select five hundred adolescents from small families and five hundred from large families, and try to discover which had experienced the greatest difficulties of adjustment at adolescence. But we could not know what were the other influences brought to bear upon these children, what effect their knowledge of sex or their neighbourhood environment may have had upon their adolescent development.

What method then is open to us who wish to conduct a human experiment but who lack the power either to construct the experimental conditions or to find controlled examples of those conditions here and there throughout our own civilisation? The only method is that of the anthropologist, to go to a different civilisation and make a study of human beings under different cultural conditions in some other part of the world. For such studies the anthropologist chooses quite simple peoples, primitive peoples, whose society has never attained the complexity of our own. In this choice of primitive peoples like the Eskimo, the Australian, the South Sea islander, or the Pueblo Indian, the anthropologist is guided by the knowledge that the analysis of a simpler civilisation is more possible of attainment.

In complicated civilisations like those of Europe, or the higher civilisations of the East, years of study are necessary before the student can begin to understand the forces at work within them. A study of the French family alone would involve a preliminary study of French history, of French law, of the Catholic and Protestant attitudes towards sex and personal relations. A primitive people without a written language present a much less elaborate problem and a trained student can master the fundamental structure of a primitive society in a few months.

Furthermore, we do not choose a simple peasant community in Europe or an isolated group of mountain whites in the American South, for these people's ways of life, though simple, belong essentially to the historical tradition to which the complex parts of European or American civilisation belong. Instead, we choose primitive groups who have had thousands of years of historical development along completely different lines from our own, whose language does not possess our Indo-European categories, whose religious ideas are of a different nature, whose social organisation is not only simpler but very different from our own. From these contrasts, which are vivid enough to startle and enlighten those accustomed to our own way of life and simple enough to be grasped quickly, it is possible to learn many things about the effect of a civilisation upon the individuals within it.

So, in order to investigate the particular problem, I chose to go not to Germany or to Russia, but to Samoa, a South Sea island about thirteen degrees from the Equator, inhabited by a brown Polynesian people. Because I was a woman and could hope for greater intimacy in working with girls rather than with boys, and because owing to a paucity of women ethnologists our knowledge of primitive girls is far slighter than our knowledge of boys, I chose to concentrate upon the adolescent girl in Samoa.

But in concentrating, I did something very different from what I would do if I concentrated upon a study of the adolescent girl in Kokomo, Indiana. In such a study, I would go right to the crux of the problem; I would not have to linger long over the Indiana language, the table manners or sleeping habits of my subjects, or make an exhaustive study of how they learned to dress themselves, to use the telephone, or what the concept of conscience meant in Kokomo. All these things are the general fabric of American life, known to me as investigator, known to you as readers.

But with this new experiment on the primitive adolescent girl the matter was quite otherwise. She spoke a language the very sounds of which were strange, a language in which nouns became verbs and verbs nouns in the most sleight-of-hand fashion. All of her habits of life were different. She sat cross-legged on the ground, and to sit upon a chair made her stiff and miserable. She ate with her fingers from a woven plate; she slept upon the floor. Her house was a mere circle of pillars, roofed by a cone of thatch, carpeted with water-worn coral fragments. Her whole material environment was different. Cocoanut palm, breadfruit, and mango trees swayed above her village. She had never seen a horse, knew no animals except the pig, dog and rat. Her food was taro, breadfruit and bananas, fish and wild pigeon and half-roasted pork, and land crabs. And just as it was necessary to understand this physical environment, this routine of life which was so different from ours, so her social environment in its attitudes towards children, towards sex, towards personality, presented as strong a contrast to the social environment of the American girl.

I concentrated upon the girls of the community. I spent the greater part of my time with them. I studied most closely the households in which adolescent girls lived. I spent more time in the games of children than in the councils of their elders. Speaking their language, eating their food, sitting barefoot and cross-legged upon the pebbly floor, I did my best to minimise the differences between us and to learn to know and understand all the girls of three little villages on the coast of the little island of Tau, in the Manu'a Archipelago.

Through the nine months which I spent in Samoa, I gathered many detailed facts about these girls, the size of their families, the position and wealth of their parents, the number of their brothers and sisters, the amount of sex experience which they had had. All of these routine facts are summarised in a table in the appendix. They are only the barest skeleton, hardly the raw materials for a study of family situations and sex relations, standards of friendship, of loyalty, of personal responsibility, all those impalpable storm centres of disturbances in the lives of our adolescent girls. And because these less measurable parts of their lives were so similar, because one girl's life was so much like another's, in an uncomplex, uniform culture like Samoa, I feel justified in generalising although I studied only fifty girls in three small neighbouring villages.

In the following chapters I have described the lives of these girls, the lives of their younger sisters who will soon be adolescent, of their brothers with whom a strict taboo forbids them to speak, of their older sisters who have left puberty behind them, of their elders, the mothers and fathers whose attitudes towards life determine the attitudes of their children. And through this description I have tried to answer the question which sent me to Samoa: Are the disturbances which vex our adolescents due to the nature of adolescence itself or to the civilisation? Under different conditions does adolescence present a different picture?

Also, by the nature of the problem, because of the unfamiliarity of this simple life on a small Pacific island, I have had to give a picture of the whole social life of Samoa, the details being selected always with a view to illuminating the problem of adolescence. Matters of political organisation which neither interest nor influence the young girl are not included. Minutiæ of relationship systems or ancestor cults, genealogies and mythology, which are of interest only to the specialist, will be published in another place. But I have tried to present to the reader the Samoan girl in her social setting, to describe the course of her life from birth until death, the problems she will have to solve, the values which will guide her in her solutions, the pains and pleasures of her human lot cast on a South Sea island.

Such a description seeks to do more than illuminate this particular problem. It should also give the reader some conception of a different and contrasting civilisation, another way of life, which other members of the human race have found satisfactory and gracious. We know that our subtlest perceptions, our highest values, are all based upon contrast; that light without darkness or beauty without ugliness would lose the qualities which they now appear to us to have. And similarly, if we would appreciate our own civilisation, this elaborate pattern of life which we have made for ourselves as a people and which we are at such pains to pass on to our children, we must set our civilisation over against other very different ones. The traveller in Europe returns to America, sensitive to nuances in his own manners and philosophies which have hitherto gone unremarked, yet Europe and America are parts of one civilisation. It is with variations within one great pattern that the student of Europe to-day or the student of our own history sharpens his sense of appreciation. But if we step outside the stream of Indo-European culture, the appreciation which we can accord our civilisation is even more enhanced. Here in remote parts of the world, under historical conditions very different from those which made Greece and Rome flourish and fall, groups of human beings have worked out patterns of life so different from our own that we cannot venture any guess that they would ever have arrived at our solutions. Each primitive people has selected one set of human gifts, one set of human values, and fashioned for themselves an art, a social organisation, a religion, which is their unique contribution to the history of the human spirit.

Samoa is only one of these diverse and gracious patterns, but as the traveller who has been once from home is wiser than he who has never left his own door step, so a knowledge of one other culture should sharpen our ability to scrutinise more steadily, to appreciate more lovingly, our own.

And, because of the particular problem which we set out to answer, this tale of another way of life is mainly concerned with education, with the process by which the baby, arrived cultureless upon the human scene, becomes a full-fledged adult member of his or her society. The strongest light will fall upon the ways in which Samoan education, in its broadest sense, differs from our own. And from this contrast we may be able to turn, made newly and vividly self-conscious and self-critical, to judge anew and perhaps fashion differently the education we give our children.

2

A Day in Samoa

THE LIFE of the day begins at dawn, or if the moon has shown until daylight, the shouts of the young men may be heard before dawn from the hillside. Uneasy in the night, populous with ghosts, they shout lustily to one another as they hasten with their work. As the dawn begins to fall among the soft brown roofs and the slender palm trees stand out against a colourless, gleaming sea, lovers slip home from trysts beneath the palm trees or in the shadow of beached canoes, that the light may find each sleeper in his appointed place. Cocks crow, negligently, and a shrill-voiced bird cries from the breadfruit trees. The insistent roar of the reef seems muted to an undertone for the sounds of a waking village. Babies cry, a few short wails before sleepy mothers give them the breast. Restless little children roll out of their sheets and wander drowsily down to the beach to freshen their faces in the sea. Boys, bent upon an early fishing, start collecting their tackle and go to rouse their more laggard companions. Fires are lit, here and there, the white smoke hardly visible against the paleness of the dawn. The whole village, sheeted and frowsy, stirs, rubs its eyes, and stumbles towards the beach. "Talofa!" "Talofa!" "Will the journey start to-day?" "Is it bonito fishing your lordship is going?" Girls stop to giggle over some young ne'er-do-well who escaped during the night from an angry father's pursuit and to venture a shrewd guess that the daughter knew more about his presence than she told. The boy who is taunted by another, who has succeeded him in his sweetheart's favour, grapples with his rival, his foot slipping in the wet sand. From the other end of the village comes a long drawn-out, piercing wail. A messenger has just brought word of the death of some relative in another village. Half-clad, unhurried women, with babies at their breasts, or astride their hips, pause in their tale of Losa's outraged departure from her father's house to the greater kindness in the home of her uncle, to wonder who is dead. Poor relatives whisper their requests to rich relatives, men make plans to set a fish trap together, a woman begs a bit of yellow dye from a kinswoman, and through the village sounds the rhythmic tattoo which calls the young men together. They gather from all parts of the village, digging sticks in hand, ready to start inland to the plantation. The older men set off upon their more lonely occupations, and each household, reassembled under its peaked roof, settles down to the routine of the morning. Little children, too hungry to wait for the late breakfast, beg lumps of cold taro which they munch greedily. Women carry piles of washing to the sea or to the spring at the far end of the village, or set off inland after weaving materials. The older girls go fishing on the reef, or perhaps set themselves to weaving a new set of Venetian blinds.

In the houses, where the pebbly floors have been swept bare with a stiff long-handled broom, the women great with child and the nursing mothers, sit and gossip with one another. Old men sit apart, unceasingly twisting palm husk on their bare thighs and muttering old tales under their breath. The carpenters begin work on the new house, while the owner bustles about trying to keep them in a good humour. Families who will cook to-day are hard at work; the taro, yams and bananas have already been brought from inland; the children are scuttling back and forth, fetching sea water, or leaves to stuff the pig. As the sun rises higher in the sky, the shadows deepen under the thatched roofs, the sand is burning to the touch, the hibiscus flowers wilt on the hedges, and little children bid the smaller ones, "Come out of the sun." Those whose excursions have been short return to the village, the women with strings of crimson jelly fish, or baskets of shell fish, the men with cocoanuts, carried in baskets slung on a shoulder pole. The women and children eat their breakfasts, just hot from the oven, if this is cook day, and the young men work swiftly in the mid-day heat, preparing the noon feast for their elders.

It is high noon. The sand burns the feet of the little children, who leave their palm leaf balls and their pin-wheels of frangipani blossoms to wither in the sun, as they creep into the shade of the houses. The women who must go abroad carry great banana leaves as sun-shades or wind wet cloths about their heads. Lowering a few blinds against the slanting sun, all who are left in the village wrap their heads in sheets and go to sleep. Only a few adventurous children may slip away for a swim in the shadow of a high rock, some industrious woman continues with her weaving, or a close little group of women bend anxiously over a woman in labour. The village is dazzling and dead; any sound seems oddly loud and out of place. Words have to cut through the solid heat slowly. And then the sun gradually sinks over the sea.

A second time, the sleeping people stir, roused perhaps by the cry of "a boat," resounding through the village. The fishermen beach their canoes, weary and spent from the heat, in spite of the slaked lime on their heads, with which they have sought to cool their brains and redden their hair. The brightly coloured fishes are spread out on the floor, or piled in front of the houses until the women pour water over them to free them from taboo. Regretfully, the young fishermen separate out the "Taboo fish," which must be sent to the chief, or proudly they pack the little palm leaf baskets with offerings of fish to take to their sweethearts. Men come home from the bush, grimy and heavy laden, shouting as they come, greeted in a sonorous rising cadence by those who have remained at home. They gather in the guest house for their evening kava drinking. The soft clapping of hands, the high-pitched intoning of the talking chief who serves the kava echoes through the village. Girls gather flowers to weave into necklaces; children, lusty from their naps and bound to no particular task, play circular games in the half shade of the late afternoon. Finally the sun sets, in a flame which stretches from the mountain behind to the horizon on the sea, the last bather comes up from the beach, children straggle home, dark little figures etched against the sky; lights shine in the houses, and each household gathers for its evening meal. The suitor humbly presents his offering, the children have been summoned from their noisy play, perhaps there is an honoured guest who must be served first, after the soft, barbaric singing of Christian hymns and the brief and graceful evening prayer. In front of a house at the end of the village, a father cries out the birth of a son. In some family circles a face is missing, in others little runaways have found a haven! Again quiet settles upon the village, as first the head of the household, then the women and children, and last of all the patient boys, eat their supper.

After supper the old people and the little children are bundled off to bed. If the young people have guests the front of the house is yielded to them. For day is the time for the councils of old men and the labours of youth, and night is the time for lighter things. Two kinsmen, or a chief and his councillor, sit and gossip over the day's events or make plans for the morrow. Outside a crier goes through the village announcing that the communal breadfruit pit will be opened in the morning, or that the village will make a great fish trap. If it is moonlight, groups of young men, women by twos and threes, wander through the village, and crowds of children hunt for land crabs or chase each other among the breadfruit trees.

Half the village may go fishing by torchlight and the curving reef will gleam with wavering lights and echo with shouts of triumph or disappointment, teasing words or smothered cries of outraged modesty. Or a group of youths may dance for the pleasure of some visiting maiden. Many of those who have retired to sleep, drawn by the merry music, will wrap their sheets about them and set out to find the dancing. A white-clad, ghostly throng will gather in a circle about the gaily lit house, a circle from which every now and then a few will detach themselves and wander away among the trees. Sometimes sleep will not descend upon the village until long past midnight; then at last there is only the mellow thunder of the reef and the whisper of lovers, as the village rests until dawn.

3

The Education of the Samoan Child

BIRTHDAYS ARE of little account in Samoa. But for the birth itself of the baby of high rank, a great feast will be held, and much property given away. The first baby must always be born in the mother's village and if she has gone to live in the village of her husband, she must go home for the occasion. For several months before the birth of the child the father's relatives have brought gifts of food to the prospective mother, while the mother's female relatives have been busy making pure white bark cloth for baby clothes and weaving dozens of tiny pandanus mats which form the layette. The expectant mother goes home laden with food gifts and when she returns to her husband's family, her family provide her with the exact equivalent in mats and bark cloth as a gift to them. At the birth itself the father's mother or sister must be present to care for the new-born baby while the midwife and the relatives of the mother care for her. There is no privacy about a birth. Convention dictates that the mother should neither writhe, nor cry out, nor inveigh against the presence of twenty or thirty people in the house who sit up all night if need be, laughing, joking, and playing games. The midwife cuts the cord with a fresh bamboo knife and then all wait eagerly for the cord to fall off, the signal for a feast. If the baby is a girl, the cord is buried under a paper mulberry tree (the tree from which bark cloth is made) to ensure her growing up to be industrious at household tasks; for a boy it is thrown into the sea that he may be a skilled fisherman, or planted under a taro plant to give him industry in farming. Then the visitors go home, the mother rises and goes about her daily tasks, and the new baby ceases to be of much interest to any one. The day, the month in which it was born, is forgotten. Its first steps or first word are remarked without exuberant comment, without ceremony. It has lost all ceremonial importance and will not regain it again until after puberty; in most Samoan villages a girl will be ceremonially ignored until she is married. And even the mother remembers only that Losa is older than Tupu, and that her sister's little boy, Fale, is younger than her brother's child, Vigo. Relative age is of great importance, for the elder may always command the younger—until the positions of adult life upset the arrangement—but actual age may well be forgotten.

Babies are always nursed, and in the few cases where the mother's milk fails her, a wet nurse is sought among the kinsfolk. From the first week they are also given other food, papaya, cocoanut milk, sugar-cane juice; the food is either masticated by the mother and then put into the baby's mouth on her finger, or if it is liquid, a piece of bark cloth is dipped into it and the child allowed to suck it, as shepherds feed orphaned lambs. The babies are nursed whenever they cry and there is no attempt at regularity. Unless a woman expects another child, she will nurse a baby until it is two or three years old, as the simplest device for pacifying its crying. Babies sleep with their mothers as long as they are at the breast; after weaning they are usually handed over to the care of some younger girl in the household. They are bathed frequently with the juice of a wild orange and rubbed with cocoanut oil until their skins glisten.

The chief nurse-maid is usually a child of six or seven who is not strong enough to lift a baby over six months old, but who can carry the child straddling the left hip, or on the small of the back. A child of six or seven months of age will assume this straddling position naturally when it is picked up. Their diminutive nurses do not encourage children to walk, as babies who can walk about are more complicated charges. They walk before they talk, but it is impossible to give the age of walking with any exactness, though I saw two babies walk whom I knew to be only nine months old, and my impression is that the average age is about a year. The life on the floor, for all activities within a Samoan house are conducted on the floor, encourages crawling, and children under three or four years of age optionally crawl or walk.

From birth until the age of four or five a child's education is exceedingly simple. They must be housebroken, a matter made more difficult by an habitual indifference to the activities of very small children. They must learn to sit or crawl within the house and never to stand upright unless it is absolutely necessary; never to address an adult in a standing position; to stay out of the sun; not to tangle the strands of the weaver; not to scatter the cut-up cocoanut which is spread out to dry; to keep their scant loin cloths at least nominally fastened to their persons; to treat fire and knives with proper caution; not to touch the kava bowl, or the kava cup; and, if their father is a chief, not to crawl on his bed-place when he is by. These are really simply a series of avoidances, enforced by occasional cuffings and a deal of exasperated shouting and ineffectual conversation.

The weight of the punishment usually falls upon the next oldest child, who learns to shout, "Come out of the sun," before she has fully appreciated the necessity of doing so herself. By the time Samoan girls and boys have reached sixteen or seventeen years of age these perpetual admonitions to the younger ones have become an inseparable part of their conversation, a monotonous, irritated undercurrent to all their comments. I have known them to intersperse their remarks every two or three minutes with, "Keep still," "Sit still," "Keep your mouths shut," "Stop that noise," uttered quite mechanically although all of the little ones present may have been behaving as quietly as a row of intimidated mice. On the whole, this last requirement of silence is continually mentioned and never enforced. The little nurses are more interested in peace than in forming the characters of their small charges and when a child begins to howl, it is simply dragged out of earshot of its elders. No mother will ever exert herself to discipline a younger child if an older one can be made responsible.

If small families of parents and children prevailed in Samoa, this system would result in making half of the population solicitous and self-sacrificing and the other half tyrannous and self-indulgent. But just as a child is getting old enough so that its wilfulness is becoming unbearable, a younger one is saddled upon it, and the whole process is repeated again, each child being disciplined and socialized through responsibility for a still younger one.

This fear of the disagreeable consequences resulting from a child's crying, is so firmly fixed in the minds of the older children that long after there is any need for it, they succumb to some little tyrant's threat of making a scene, and five-year-olds bully their way into expeditions on which they will have to be carried, into weaving parties where they will tangle the strands, and cook houses where they will tear up the cooking leaves or get thoroughly smudged with the soot and have to be washed—all because an older boy or girl has become so accustomed to yielding any point to stop an outcry.

This method of giving in, coaxing, bribing, diverting the infant disturbers is only pursued within the household or the relationship group, where there are duly constituted elders in authority to punish the older children who can't keep the babies still. Towards a neighbour's children or in a crowd the half-grown girls and boys and even the adults vent their full irritation upon the heads of troublesome children. If a crowd of children are near enough, pressing in curiously to watch some spectacle at which they are not wanted, they are soundly lashed with palm leaves, or dispersed with a shower of small stones, of which the house floor always furnishes a ready supply. This treatment does not seem actually to improve the children's behaviour, but merely to make them cling even closer to their frightened and indulgent little guardians. It may be surmised that stoning the children from next door provides a most necessary outlet for those who have spent so many weary hours placating their own young relatives. And even these bursts of anger are nine-tenths gesture. No one who throws the stones actually means to hit a child, but the children know that if they repeat their intrusions too often, by the laws of chance some of the flying bits of coral will land in their faces. Even Samoan dogs have learned to estimate the proportion of gesture that there is in a Samoan's "get out of the house." They simply stalk out between one set of posts and with equal dignity and all casualness stalk in at the next opening.

By the time a child is six or seven she has all the essential avoidances well enough by heart to be trusted with the care of a younger child. And she also develops a number of simple techniques. She learns to weave firm square balls from palm leaves, to make pin-wheels of palm leaves or frangipani blossoms, to climb a cocoanut tree by walking up the trunk on flexible little feet, to break open a cocoanut with one firm well-directed blow of a knife as long as she is tall, to play a number of group games and sing the songs which go with them, to tidy the house by picking up the litter on the stony floor, to bring water from the sea, to spread out the copra to dry and to help gather it in when rain threatens, to roll the pandanus leaves for weaving, to go to a neighbouring house and bring back a lighted fagot for the chief's pipe or the cookhouse fire, and to exercise tact in begging slight favours from relatives.

But in the case of the little girls all of these tasks are merely supplementary to the main business of baby-tending. Very small boys also have some care of the younger children, but at eight or nine years of age they are usually relieved of it. Whatever rough edges have not been smoothed off by this responsibility for younger children are worn off by their contact with older boys. For little boys are admitted to interesting and important activities only so long as their behaviour is circumspect and helpful. Where small girls are brusquely pushed aside, small boys will be patiently tolerated and they become adept at making themselves useful. The four or five little boys who all wish to assist at the important business of helping a grown youth lasso reef eels, organize themselves into a highly efficient working team; one boy holds the bait, another holds an extra lasso, others poke eagerly about in holes in the reef looking for prey, while still another tucks the captured eels into his *lavalava*. The small girls, burdened with heavy babies or the care of little staggerers who are too small to adventure on the reef, discouraged by the hostility of the small boys and the scorn of the older ones, have little opportunity for learning the more adventurous forms of work and play. So while the little boys first undergo the chastening effects of baby-tending and then have many opportunities to learn effective co-operation under the supervision of older boys, the girls' education is less comprehensive. They have a high standard of individual responsibility but the community provides them with no lessons in co-operation with one another. This is particularly apparent in the activities of young people; the boys organise quickly; the girls waste hours in bickering, innocent of any technique for quick and efficient co-operation.

And as the woman who goes fishing can only get away by turning the babies over to the little girls of the household, the little girls cannot accompany their aunts and mothers. So they learn even the simple processes of reef fishing much later than do the boys. They are kept at the baby-tending, errand-running stage until they are old enough and robust enough to work on the plantations and carry foodstuffs down to the village.

A girl is given these more strenuous tasks near the age of puberty, but it is purely a question of her physical size and ability to take responsibility, rather than of her physical maturity. Before this time she has occasionally accompanied the older members of the family to the plantations if they were willing to take the babies along also. But once there, while her brothers and cousins are collecting cocoanuts and roving happily about in the bush, she has again to chase and shepherd and pacify the ubiquitous babies.

As soon as the girls are strong enough to carry heavy loads, it pays the family to shift the responsibility for the little children to the younger girls and the adolescent girls are released from baby-tending. It may be said with some justice that the worst period of their lives is over. Never again will they be so incessantly at the beck and call of their elders, never again so tyrannised over by two-year-old tyrants. All the irritating, detailed routine of housekeeping, which in our civilisation is accused of warping the souls and souring the tempers of grown women, is here performed by children under fourteen years of age. A fire or a pipe to be kindled, a call for a drink, a lamp to be lit, the baby's cry, the errand of the capricious adult—these haunt them from morning until night. With the introduction of several months a year of government schools these children are being taken out of their homes for most of the day. This brings about a complete disorganisation of the native households which have no precedents for a manner of life where mothers have to stay at home and take care of their children and adults have to perform small routine tasks and run errands.

Before their release from baby-tending the little girls have a very limited knowledge of any of the more complicated techniques. Some of them can do the simpler work in preparing food for cooking, such as skinning bananas, grating cocoanuts or scraping taro. A few of them can weave the simple carrying basket. But now they must learn to weave all their own baskets for carrying supplies; learn to select taro leaves of the right age for cooking, to dig only mature taro. In the cook-house they learn to make *palusami*, to grate the cocoanut meat, season it with hot stones, mix it with sea water and strain out the husks, pour this milky mixture into a properly made little container of taro leaves from which the aromatic stem has been scorched off, wrap these in a breadfruit leaf and fasten the stem tightly to make a durable cooking jacket. They must learn to lace a large fish into a palm leaf, or roll a bundle of small fish in a breadfruit leaf; to select the right kind of leaves for stuffing a pig, to judge when the food in the oven of small heated stones is thoroughly baked. Theoretically the bulk of the cooking is done by the boys and where a girl has to do the heavier work, it is a matter for comment: "Poor Losa, there are no boys in her house and always she must make the oven." But the girls always help and often do a great part of the work.

Once they are regarded as individuals who can devote a long period of time to some consecutive activity, girls are sent on long fishing expeditions. They learn to weave fish baskets, to gather and arrange the bundles of fagots used in torch-light fishing, to tickle a devil fish until it comes out of its hole and climbs obediently upon the waiting stick, appropriately dubbed a "come hither stick"; to string the great rose-coloured jellyfish, *lole,* a name which Samoan children give to candy also, on a long string of hibiscus bark, tipped with a palm leaf rib for a needle; to know good fish from bad fish, fish that are in season from fish which are dangerous at some particular time of the year; and never to take two octopuses, found paired on a rock, lest bad luck come upon the witless fisher.

Before this time their knowledge of plants and trees is mainly a play one, the pandanus provides them with seeds for necklaces, the palm tree with leaves to weave balls; the banana tree gives leaves for umbrellas and half a leaf to shred into a stringy "choker"; cocoanut shells cut in half, with cinet strings attached, make a species of stilt; the blossoms of the *Pua* tree can be sewed into beautiful necklaces. Now they must learn to recognise these trees and plants for more serious purposes; they must learn when the pandanus leaves are ready for the cutting and how to cut the long leaves with one sure quick stroke; they must distinguish between the three kinds of pandanus used for different grades of mats. The pretty orange seeds which made such attractive and also edible necklaces must now be gathered as paint brushes for ornamenting bark cloth. Banana leaves are gathered to protect the woven platters, to wrap up puddings for the oven, to bank the steaming oven full of food. Banana bark must be stripped at just the right point to yield the even, pliant, black strips, needed to ornament mats and baskets. Bananas themselves must be distinguished as to those which are ripe for burying, or the golden curved banana ready for eating, or bananas ready to be sun-dried for making fruit-cake rolls. Hibiscus bark can no longer be torn off at random to give a raffia-like string for a handful of shells; long journeys must be made inland to select bark of the right quality for use in weaving.

In the house the girl's principal task is to learn to weave. She has to master several different techniques. First, she learns to weave palm branches where the central rib of the leaf serves as a rim to her basket or an edge to her mat and where the leaflets are already arranged for weaving. From palm leaves she first learns to weave a carrying basket, made of half a leaf, by plaiting the leaflets together and curving the rib into a rim. Then she learns to weave the Venetian blinds which hang between the house posts, by laying one-half leaf upon another and plaiting the leaflets together. More difficult are the floor mats, woven of four great palm leaves, and the food platters with their intricate designs. There are also fans to make, simple two-strand weaves which she learns to make quite well, more elaborate twined ones which are the prerogative of older and more skilled weavers. Usually some older woman in the household trains a girl to weave and sees to it that she makes at least one of each kind of article, but she is only called upon to produce in quantity the simpler things, like the Venetian blinds. From the pandanus she learns to weave the common floor mats, one or two types of the more elaborate bed mats, and then, when she is thirteen or fourteen, she begins her first fine mat. The fine mat represents the high point of Samoan weaving virtuosity. Woven of the finest quality of pandanus which has been soaked and baked and scraped to a golden whiteness and paper-like thinness, of strands a sixteenth of an inch in width, these mats take a year or two years to weave and are as soft and pliable as linen. They form the unit of value, and must always be included in the dowry of the bride. Girls seldom finish a fine mat until they are nineteen or twenty, but the mat has been started, and, wrapped up in a coarser one, it rests among the rafters, a testimony to the girl's industry and manual skill. She learns the rudiments of bark cloth making; she can select and cut the paper mulberry wands, peel off the bark, beat it after it has been scraped by more expert hands. The patterning of the cloth with a pattern board or by free hand drawing is left for the more experienced adult.

Throughout this more or less systematic period of education, the girls maintain a very nice balance between a reputation for the necessary minimum of knowledge and a virtuosity which would make too heavy demands. A girl's chances of marriage are badly damaged if it gets about the village that she is lazy and inept in domestic tasks. But after these first stages have been completed the girl marks time technically for three or four years. She does the routine weaving, especially of the Venetian blinds and carrying baskets. She helps with the plantation work and the cooking, she weaves a very little on her fine mat. But she thrusts virtuosity away from her as she thrusts away every other sort of responsibility with the invariable comment, "Laititi a'u" (I am but young). All of her interest is expended on clandestine sex adventures, and she is content to do routine tasks as, to a certain extent, her brother is also.

But the seventeen-year-old boy is not left passively to his own devices. He has learned the rudiments of fishing, he can take a dug-out canoe over the reef safely, or manage the stern paddle in a bonito boat. He can plant taro or transplant cocoanut, husk cocoanuts on a stake and cut the meat out with one deft quick turn of the knife. Now at seventeen or eighteen he is thrust into the *Aumaga*, the society of the young men and the older men without titles, the group that is called, not in euphuism but in sober fact, "the strength of the village." Here he is badgered into efficiency by rivalry, precept and example. The older chiefs who supervise the activities of the *Aumaga* gaze equally sternly upon any back-slidings and upon any undue precocity. The prestige of his group is ever being called into account by the *Aumaga* of the neighbouring villages. His fellows ridicule and persecute the boy who fails to appear when any group activity is on foot, whether work for the village on the plantations, or fishing, or cooking for the chiefs, or play in the form of a ceremonial call upon some visiting maiden. Furthermore, the youth is given much more stimulus to learn and also a greater variety of occupations are open to him. There is no specialisation among women, except in medicine and mid-wifery, both the prerogatives of very old women who teach their arts to their middle-aged daughters and nieces. The only other vocation is that of the wife of an official orator, and no girl will prepare herself for this one type of marriage which demands special knowledge, for she has no guarantee that she will marry a man of this class.

For the boy it is different. He hopes that some day he will hold a *matai* name, a name which will make him a member of the *Fono*, the assembly of headmen, which will give him a right to drink kava with chiefs, to work with chiefs rather than with young men, to sit inside the house, even though his new title is only of "between the posts" rank, and not of enough importance to give him a right to a post for his back. But very seldom is he absolutely assured of getting such a name. Each family holds several of these titles which they confer upon the most promising youths in the whole family connection. He has many rivals. They also are in the *Aumaga*. He must always pit himself against them in the group activities. There are also several types of activities in one of which he must specialise. He must become a house-builder, a fisherman, an orator or a wood carver. Proficiency in some technique must set him off a little from his fellows. Fishing prowess means immediate rewards in the shape of food gifts to offer to his sweetheart; without such gifts his advances will be scorned. Skill in house-building means wealth and status, for a young man who is a skilled carpenter must be treated as courteously as a chief and addressed with the chief's language, the elaborate set of honorific words used to people of rank. And with this goes the continual demand that he should not be too efficient, too outstanding, too precocious. He must never excel his fellows by more than a little. He must neither arouse their hatred nor the disapproval of his elders who are far readier to encourage and excuse the laggard than to condone precocity. And at the same time he shares his sister's reluctance to accept responsibility, and if he should excel gently, not too obviously, he has good chances of being made a chief. If he is sufficiently talented, the *Fono* itself may deliberate, search out a vacant title to confer upon him and call him in that he may sit with the old men and learn wisdom. And yet so well recognised is the unwillingness of the young men to respond to this honour, that the provision is always made, "And if the young man runs away, then never shall he be made a chief, but always he must sit outside the house with the young men, preparing and serving the food of the *matais* with whom he may not sit in the *Fono.*" Still more pertinent are the chances of his relationship group bestowing a *matai* name upon the gifted young man. And a *matai* he wishes to be, some day, some far-off day when his limbs have lost a little of their suppleness and his heart the love of fun and of dancing. As one chief of twenty-seven told me: "I have been a chief only four years and look, my hair is grey, although in Samoa grey hair comes very slowly, not in youth, as it comes to the white man. But always, I must act as if I were old. I must walk gravely and with a measured step. I may not dance except upon most solemn occasions, neither may I play games with the young men. Old men of sixty are my companions and watch my every word, lest I make a mistake. Thirty-one people live in my household. For them I must plan, I must find them food and clothing, settle their disputes, arrange their marriages. There is no one in my whole family who dares to scold me or even to address me familiarly by my first name. It is hard to be so young and yet to be a chief." And the old men shake their heads and agree that it is unseemly for one to be a chief so young.

The operation of natural ambition is further vitiated by the fact that the young man who is made a *matai* will not be the greatest among his former associates, but the youngest and greenest member of the *Fono*. And no longer may he associate familiarly with his old companions; a *matai* must associate only with *matais,* must work beside them in the bush and sit and talk quietly with them in the evening.

And so the boy is faced by a far more difficult dilemma than the girl. He dislikes responsibility, but he wishes to excel in his group; skill will hasten the day when he is made a chief, yet he receives censure and ridicule if he slackens his efforts; but he will be scolded if he proceeds too rapidly; yet if he would win a sweetheart, he must have prestige among his fellows. And conversely, his social prestige is increased by his amorous exploits.

So while the girl rests upon her "pass" proficiency, the boy is spurred to greater efforts. A boy is shy of a girl who does not have these proofs of efficiency and is known to be stupid and unskilled; he is afraid he may come to want to marry her. Marrying a girl without proficiency would be a most imprudent step and involve an endless amount of wrangling with his family. So the girl who is notoriously inept must take her lovers from among the casual, the jaded, and the married who are no longer afraid that their senses will betray them into an imprudent marriage.

But the seventeen-year-old girl does not wish to marry—not yet. It is better to live as a girl with no responsibility, and a rich variety of emotional experience. This is the best period of her life. There are as many beneath her whom she may bully as there are others above her to tyrannise over her. What she loses in prestige, she gains in freedom. She has very little baby-tending to do. Her eyes do not ache from weaving nor does her back break from bending all day over the tapa board. The long expeditions after fish and food and weaving materials give ample opportunities for rendezvous. Proficiency would mean more work, more confining work, and earlier marriage, and marriage is the inevitable to be deferred as long as possible.

4

The Samoan Household

A SAMOAN VILLAGE is made up of some thirty to forty households, each of which is presided over by a headman called a *matai*. These headmen hold either chiefly titles or the titles of talking chiefs, who are the official orators, spokesmen and ambassadors of chiefs. In a formal village assembly each *matai* has his place, and represents and is responsible for all the members of his household. These households include all the individuals who live for any length of time under the authority and protection of a common *matai*. Their composition varies from the biological family consisting of parents and children only, to households of fifteen and twenty people who are all related to the *matai* or to his wife by blood, marriage or adoption, but who often have no close relationship to each other. The adopted members of a household are usually but not necessarily distant relatives.

Widows and widowers, especially when they are childless,

usually return to their blood relatives, but a married couple may live with the relatives of either one. Such a household is not necessarily a close residential unit, but may be scattered over the village in three or four houses. No one living permanently in another village is counted as a member of the household, which is strictly a local unit. Economically, the household is also a unit, for all work upon the plantations is under the supervision of the *matai* who in turn parcels out to them food and other necessities.

Within the household, age rather than relationship gives disciplinary authority. The *matai* exercises nominal and usually real authority over every individual under his protection, even over his father and mother. This control is, of course, modified by personality differences, always carefully tempered, however, by a ceremonious acknowledgment of his position. The newest baby born into such a household is subject to every individual in it, and his position improves no whit with age until a younger child appears upon the scene. But in most households the position of youngest is a highly temporary one. Nieces and nephews or destitute young cousins come to swell the ranks of the household and at adolescence a girl stands virtually in the middle with as many individuals who must obey her as there are persons to whom she owes obedience. Where increased efficiency and increased self-consciousness would perhaps have made her obstreperous and restless in a differently organised family, here she has ample outlet for a growing sense of authority.

This development is perfectly regular. A girl's marriage makes a minimum of difference in this respect, except in so far as her own children increase most pertinently the supply of agreeably docile subordinates. But the girls who remain unmarried even beyond their early twenties are in nowise less highly regarded or less responsible than their married sisters. This tendency to make the classifying principle age, rather than married state, is reinforced outside the home by the fact that the wives of untitled men and all unmarried girls past puberty are classed together in the ceremonial organisation of the village.

Relatives in other households also play a rôle in the children's lives. Any older relative has a right to demand personal service from younger relatives, a right to criticise their conduct and to interfere in their affairs. Thus a little girl may escape alone down to the beach to bathe only to be met by an older cousin who sets her washing or caring for a baby or to fetch some cocoanut to scrub the clothes. So closely is the daily life bound up with this universal servitude and so numerous are the acknowledged relationships in the name of which service can be exacted, that for the children an hour's escape from surveillance is almost impossible.

This loose but demanding relationship group has its compensation, also. Within it a child of three can wander safely and come to no harm, can be sure of finding food and drink, a sheet to wrap herself up in for a nap, a kind hand to dry casual tears and bind up her wounds. Any small children who are missing when night falls, are simply "sought among their kinsfolk," and a baby whose mother has gone inland to work on the plantation is passed from hand to hand for the length of the village.

The ranking by age is disturbed in only a few cases. In each village one or two high chiefs have the hereditary right to name some girl of their household as its *taupo,* the ceremonial princess of the house. The girl who at fifteen or sixteen is made a *taupo* is snatched from her age group and sometimes from her immediate family also and surrounded by a glare of prestige. The older women of the village accord her courtesy titles, her immediate family often exploits her position for their personal ends and in return show great consideration for her wishes. But as there are only two or three *taupos* in a village, their unique position serves to emphasise rather than to disprove the general status of young girls.

Coupled with this enormous diffusion of authority goes a fear of overstraining the relationship bond, which expresses itself in an added respect for personality. The very number of her captors is the girl's protection, for does one press her too far, she has but to change her residence to the home of some more complacent relative. It is possible to classify the different households open to her as those with hardest work, least chaperonage, least scolding, largest or least number of contemporaries, fewest babies, best food, etc. Few children live continuously in one household, but are always testing out other possible residences. And this can be done under the guise of visits and with no suggestion of truancy. But the minute that the mildest annoyance grows up at home, the possibility of flight moderates the discipline and alleviates the child's sense of dependency. No Samoan child, except the *taupo,* or the thoroughly delinquent, ever has to deal with a feeling of being trapped. There are always relatives to whom one can flee. This is the invariable answer which a Samoan gives when some familial impasse is laid before him. "But she will go to some *other* relative." And theoretically the supply of relatives is inexhaustible. Unless the vagrant has committed some very serious offence like incest, it is only necessary formally to depart from the bosom of one's household. A girl whose father has beaten her over severely in the morning will be found living in haughty sanctuary, two hundred feet away, in a different household. So cherished is this system of consanguineous refuge, that an untitled man or a man of lesser rank will beard the nobler relative who comes to demand a runaway child. With great politeness and endless expressions of conciliation, he will beg his noble chief to return to his noble home and remain there quietly until his noble anger is healed against his noble child.

The most important relationships* within a Samoan household which influence the lives of the young people are the relationships between the boys and girls who call each other "brother" and "sister," whether by blood, marriage or adoption and the relationship between younger and older relatives. The stress upon the sex difference between contemporaries and the emphasis on relative age are amply explained by the conditions of family life. Relatives of opposite sex have a most rigid code of etiquette prescribed for all their contacts with each other. After they have reached years of discretion, nine or ten years of age in this case, they may not touch each other, sit close together, eat together, address each other familiarly, or mention any salacious matter in each other's presence. They may not remain in any house, except their own, together, unless half the village is gathered there. They may not walk together, use each other's possessions, dance on the same floor, or take part in any of the same small group activities. This strict avoidance applies to all individuals of the opposite sex within five years above or below one's own age with whom one was reared or to whom one acknowledges relationship by blood or marriage. The conformance to this brother and sister taboo begins when the younger of the two children feels "ashamed" at the elder's touch and continues until old age when the decrepit, toothless pair of old siblings may again sit on the same mat and not feel ashamed.

Tei, the word for younger relative, stresses the other most emotionally charged relationship. The first maternal enthusiasm of a girl is never expended upon her own children but upon some younger relative. And it is the girls and women who use this term most, continuing to cherish it after they and the younger ones to whom it is applied are full grown. The younger child in turn expends its enthusiasm upon a still younger one without manifesting any excessive affection for the fostering elders.

The word *aiga* is used roughly to cover all relationships

* See Appendix, page 146.

by blood, marriage and adoption, and the emotional tone seems to be the same in each case. Relationship by marriage is counted only as long as an actual marriage connects two kinship groups. If the marriage is broken in any way, by desertion, divorce, or death, the relationship is dissolved and members of the two families are free to marry each other. If the marriage left any children, a reciprocal relationship exists between the two households as long as the child lives, for the mother's family will always have to contribute one kind of property, the father's family another, for occasions when property must be given away in the name of the child.

A relative is regarded as some one upon whom one has a multitude of claims and to whom one owes a multitude of obligations. From a relative one may demand food, clothing, and shelter, or assistance in a feud. Refusal of such a demand brands one as stingy and lacking in human kindness, the virtue most esteemed among the Samoans. No definite repayment is made at the time such services are given, except in the case of the distribution of food to all those who share in a family enterprise. But careful count of the value of the property given and of the service rendered is kept and a return gift demanded at the earliest opportunity. Nevertheless, in native theory the two acts are separate, each one in turn becoming a "beggar," a pensioner upon another's bounty. In olden times, the beggar sometimes wore a special girdle which delicately hinted at the cause of his visit. One old chief gave me a graphic description of the behaviour of some one who had come to ask a favour of a relative. "He will come early in the morning and enter quietly, sitting down in the very back of the house, in the place of least honour. You will say to him, 'So you have come, be welcome!' and he will answer, 'I have come indeed, saving your noble presence.' Then you will say, 'Are you thirsty? Alas for your coming, there is little that is good within the house.' And he will answer, 'Let it rest, thank you, for indeed I am not hungry nor would I drink.' And he will sit and you will sit all day long and no mention is made of the purpose of his coming. All day he will sit and brush the ashes out of the hearth, performing this menial and dirty task with very great care and attention. If some one must go inland to the plantation to fetch food, he is the first to offer to go. If some one must go fishing to fill out the crew of a canoe, surely he is delighted to go, even though the sun is hot and his journey hither has been long. And all day you sit and wonder, 'What can it be that he has come for? Is it that largest pig that he wants, or has he heard perhaps that my daughter has just finished a large and beautiful piece of tapa? Would it perhaps be well to send that tapa, as I had perhaps planned, as a present to my talking chief, to send it now, so that I may refuse him with all good faith?' And he sits and studies your countenance and wonders if you will be favourable to his request. He plays with the children but refuses the necklace of flowers which they have woven for him and gives it instead to your daughter. Finally night comes. It is time to sleep and still he has not spoken. So finally you say to him, 'Lo, I would sleep. Will you sleep also or will you be returning whence you have come?' And only then will he speak and tell you the desire in his heart."

So the intrigue, the needs, the obligations of the larger relationship group which threads its carefully remembered way in and out of many houses and many villages, cuts across the life of the household. One day it is the wife's relatives who come to spend a month or borrow a fine mat; the next day it is the husband's; the third, a niece who is a valued worker in the household may be called home by the illness of her father. Very seldom do all of even the small children of a biological family live in one household and while the claims of the household are paramount, in the routine of everyday life, illness or need on the part of the closer relative in another household will call the wanderers home again.

Obligations either to give general assistance or to give specific traditionally required service, as in a marriage or at a birth, follow relationship lines, not household lines. But a marriage of many years' duration binds the relationship groups of husband and wife so closely together that to all appearances it is the household unit which gives aid and accedes to a request brought by the relative of either one. Only in families of high rank where the distaff side has priority in decisions and in furnishing the *taupo*, the princess of the house, and the male line priority in holding the title, does the actual blood relationship continue to be a matter of great practical importance; and this importance is lost in the looser household group constituted as it is by the three principles of blood, marriage and adoption, and bound together by common ties of everyday living and mutual economic dependence.

The *matai* of a household is theoretically exempt from the performance of small domestic tasks, but he is seldom actually so except in the case of a chief of high rank. However, the leading role is always accorded to him in any industrial pursuit; he dresses the pig for the feasts and cuts up the cocoanuts which the boys and women have gathered. The family cooking is done by the men and women both, but the bulk of the work falls upon the boys and young men.

The old men spin the cocoanut fibre, and braid it into the native cord which is used for fish lines, fish nets, to sew canoe parts together and to bind all the different parts of a house in place. With the old women who do the bulk of the weaving and making of bark cloth, they supervise the younger children who remain at home. The heavy routine agricultural work falls upon the women who are responsible for the weeding, transplanting, gathering and transportation of the food, and the gathering of the paper mulberry wands from which bark will be peeled for making tapa, of the hibiscus bark and pandanus leaves for weaving mats. The older girls and women also do the routine reef fishing for octopuses, sea eggs, jelly fish, crabs, and other small fry. The younger girls carry the water, care for the lamps (to-day, except in times of great scarcity when the candle nut and cocoanut oil are resorted to, the natives use kerosene lamps and lanterns), and sweep and arrange the houses. Tasks are all graduated with a fair recognition of abilities which differ with age, and, except in the case of individuals of very high rank, a task is rejected because a younger person has skill enough to perform it, rather than because it is beneath an adult's dignity.

Rank in the village and rank in the household reflect each other, but village rank hardly affects the young children. If a girl's father is a *matai*, the *matai* of the household in which she lives, she has no appeal from his authority. But if some other member of the family is the *matai*, he and his wife may protect her from her father's exactions. In the first case, disagreement with her father means leaving the household and going to live with other relatives; in the second case it may mean only a little internal friction. Also in the family of a high chief or a high talking chief there is more emphasis upon ceremonial, more emphasis upon hospitality. The children are better bred and also much harder worked. But aside from the general quality of a household which is dependent upon the rank of its head, households of very different rank may seem very similar to young children. They are usually more concerned with the temperament of those in authority than with their rank. An uncle in another village who is a very high chief is of much less significance in a child's life than some old woman in her own household who has a frightful temper.

Nevertheless, rank not of birth but of title is very important in Samoa. The status of a village depends upon the rank of its high chief, the prestige of a household depends upon the title of its *matai*. Titles are of two grades, chiefs and talking chiefs; each title carries many other duties and prerogatives besides the headship of a household. And the Samoans find rank a never-failing source of interest. They have invented an elaborate courtesy language which must be used to people of rank; complicated etiquette surrounds each rank in society. Something which concerns their elders so nearly cannot help being indirectly reflected in the lives of some of the children. This is particularly true of the relationship of children to each other in households which hold titles to which some of them will one day attain. How these faraway issues of adult life effect the lives of children and young people can best be understood by following their influence in the lives of particular children.

In the household of a high chief named Malae lived two little girls, Meta, twelve, and Timu, eleven. Meta was a self-possessed, efficient little girl. Malae had taken her from her mother's house—her mother was his cousin—because she showed unusual intelligence and precocity. Timu, on the other hand, was an abnormally shy, backward child, below her age group in intelligence. But Meta's mother was only a distant cousin of Malae. Had she not married into a strange village where Malae was living temporarily, Meta might never have come actively to the notice of her noble relative. And Timu was the only daughter of Malae's dead sister. Her father had been a quarter caste which served to mark her off and increase her self-consciousness. Dancing was an agony to her. She fled precipitately from an elder's admonitory voice. But Timu would be Malae's next *taupo*, princess. She was pretty, the principal recognised qualification, and she came from the distaff side of the house, the preferred descent for a *taupo*. So Meta, the more able in every way, was pushed to the wall, and Timu, miserable over the amount of attention she received, was dragged forward. The mere presence of another more able and enterprising child would probably have emphasised Timu's feeling of inferiority, but this publicity stressed it painfully. Commanded to dance on every occasion, she would pause whenever she caught an onlooker's eye and stand a moment wringing her hands before going on with the dance.

In another household, this same title of Malae's *taupo* played a different role. This was in the household of Malae's paternal aunt who lived with her husband in Malae's guest house in his native village. Her eldest daughter, Pana, held the title of *taupo* of the house of Malae. But Pana was twenty-six, though still unmarried. She must be wedded soon and then another girl must be found to hold the title. Timu would still be too young. Pana had three younger sisters who by birth were supremely eligible to the title. But Mele, the eldest of twenty, was lame, and Pepe of fourteen was blind in one eye and an incorrigible tomboy. The youngest was even younger than Timu. So all three were effectually barred from succession. This fact reacted favourably upon the position of Filita. She was a seventeen-year-old niece of the father of the other children with no possible claims on a title in the house of Malae, but she had lived with her cousins since childhood. Filita was pretty, efficient, adequate, neither lame like Mele nor blind and hoydenish like Pepe. True she could never hope to be *taupo*, but neither could they, despite their superior birth, so peace and amity reigned because of her cousins' deficiencies. Still another little girl came within the circle of influence of the title. This was Pula, another little cousin in a third village. But her more distant relationship and possible claims were completely obscured by the fact that she was the only granddaughter of the highest chief in her own village and her becoming the *taupo* of that title was inevitable so that her life was untouched by any other possibility. Thus six girls in addition to the present *taupo*, were influenced for good or evil by the possibility of succession to the title. But as there are seldom more than one or two *taupos* in a village, these influences are still fairly circumscribed when compared with the part which rank plays in the lives of boys, for there are usually one or more *matai* names in every relationship group.

Rivalry plays a much stronger part here. In the choice of the *taupo* and the *manaia* (the titular heir-apparent) there is a strong prejudice in favour of blood relationship and also for the choice of the *taupo* from the female and the *manaia* from the male line. But in the interests of efficiency this scheme had been modified, so that most titles were filled by the most able youth from the whole relationship and affinity group. So it was in Alofi. Tui, a chief of importance in the village, had one son, an able intelligent boy. Tui's brothers were dull and inept, no fit successors to the title. One of them had an ill-favoured young son, a stupid, unattractive youngster. There were no other males in the near relationship group. It was assumed that the exceedingly eligible son would succeed his father. And then at twenty he died. The little nephew hardly gave promise of a satisfactory development, and so Tui had his choice of looking outside his village or outside of his near relationship group. Village feeling runs high in Tui's village. Tui's blood relatives lived many villages away. They were strangers. If he did not go to them and search for a promising youth whom he could train as his successor, he must either find an eligible young husband for his daughter or look among his wife's people.

Provisionally he took this last course, and his wife's brother's son came to live in his household. In a year, his new father promised the boy, he might assume his dead cousin's name if he showed himself worthy.

In the family of high chief Fua a very different problem presented itself. His was the highest title in the village. He was over sixty and the question of succession was a moot one. The boys in his household consisted of Tata, his eldest son who was illegitimate, Molo and Nua, the sons of his widowed sister, Sisi, his son by his first legal wife (since divorced and remarried on another island), and Tuai, the husband of his niece, the sister of Molo and Nua. And in the house of Fua's eldest brother lived his brother's daughter's son, Alo, a youth of great promise. Here then were enough claimants to produce a lively rivalry. Tuai was the oldest, calm, able, but not sufficiently hopeful to be influenced in his conduct except as it made him more ready to assert the claims of superior age over his wife's younger brothers whose claims were better than his. Next in age came Tata, the sour, beetle-browed bastard, whose chances were negligible as long as there were those of legitimate birth to dispute his left-hand claims. But Tata did not lose hope. Cautious, tortuous-minded, he watched and waited. He was in love with Lotu, the daughter of a talking chief of only medium rank. For one of Fua's sons, Lotu would have been a good match. But as Fua's bastard who wished to be chief, he must marry high or not at all. The two nephews, Molo and Nua, played different hands. Nua, the younger, went away to seek his fortune as a native marine at the Naval Station. This meant a regular income, some knowledge of English, prestige of a sort. Molo, the elder brother, stayed at home and made himself indispensable. He was the *tamafafine*, the child of the distaff side, and it was his rôle to take his position for granted, the *tamafafine* of the house of Fua, what more could any one ask in the way of immediate prestige. As for the future—his manner was perfect. All of these young men, and likewise Alo, the great-nephew, were members of the *Aumaga*, grown up and ready to assume adult responsibilities. Sisi, the sixteen-year-old legitimate son, was still a boy, slender, diffident, presuming far less upon his position as son and heir-apparent than did his cousin. He was an attractive, intelligent boy. If his father lived until Sisi was twenty-five or thirty, his succession seemed inevitable. Even should his father die sooner, the title might have been held for him. But in this latter possibility there was one danger. Samala, his father's older brother, would have a strong voice in the choice of a successor to the title. And Alo was Samala's adored grandson, the son of his favourite daughter. Alo was the model of all that a young man should be. He eschewed the company of women, stayed much at home and rigorously trained his younger brother and sister. While the other young men played cricket, he sat at Samala's feet and memorised genealogies. He never forgot that he was the son of Sāfuá, the house of Fua. More able than Molo, his claim to the title was practically as good, although within the family group Molo as the child of the distaff side would always outvote him. So Alo was Sisi's most dangerous rival, provided his father died soon. And should Fua live twenty years longer, another complication threatened his succession. Fua had but recently re-married, a woman of very high rank and great wealth who had a five-year-old illegitimate son, Nifo. Thinking always of this child, for she and Fua had no children, she did all that she could to undermine Sisi's position as heir-apparent and there was every chance that as her ascendency over Fua increased with his advancing age, she might have Nifo named as his successor. His illegitimacy and lack of blood tie would be offset by the fact that he was child of the distaff side in the noblest family in the island and would inherit great wealth from his mother.

Of a different character was the problem which confronted Sila, the stepdaughter of Ono, a *matai* of low rank. She was the eldest in a family of seven children. Ono was an old man, decrepit and ineffective. Lefu, Sila's mother and his second wife, was worn out, weary from bearing eleven children. The only adult males in the household were Laisa, Ono's brother, an old man like himself, and Laisa's idle shiftless son, a man of thirty, whose only interest in life was love affairs. He was unmarried and shied away from this responsibility as from all others. The sister next younger than Sila was sixteen. She had left home and lived, now here, now there, among her relatives. Sila was twenty-two. She had been married at sixteen and against her will to a man much older than herself who had beaten her for her childish ways. After two years of married life, she had run away from her husband and gone home to live with her parents, bringing her little two-year-old boy, who was now five years old, with her. At twenty she had a love affair with a boy of her own village, and borne a daughter who had lived only a few months. After her baby died her lover had deserted her. Sila disliked matrimony. She was conscientious, sharp-tongued, industrious. She worked tirelessly for her child and her small brothers and sisters. She did not want to marry again. But there were three old people and six children in her household with only herself and her idle cousin to provide for them.

And so she said despondently: "I think I will get married to that boy." "Which boy, Sila?" I asked. "The father of my baby who is dead." "But I thought you said you did not want him for a husband?" "No more do I. But I must find some one to care for my family." And indeed there was no other way. Her stepfather's title was a very low one. There were no young men within the family to succeed to it. Her lover was industrious and of even lower degree. The bait of the title would secure a worker for the family.

And so within many households the shadow of nobility falls upon the children, sometimes lightly, sometimes heavily, often long before they are old enough to understand the meaning of these intrusions from the adult world.

5

The Girl and Her Age Group

UNTIL A CHILD is six or seven at least she associates very little with her contemporaries. Brothers and sisters and small cousins who live in the same household, of course, frolic and play together, but outside the household each child clings closely to its older guardian and only comes in contact with other children in case the little nursemaids are friends. But at about seven years of age, the children begin to form larger groups, a kind of voluntary association which never exists in later life, that is, a group recruited from both relationship and neighbourhood groups. These are strictly divided along sex lines and antagonism between the small girls and the small boys is one of the salient features of the group life. The little girls are just beginning to "be ashamed" in the presence of older brothers, and the prohibition that one small girl must never join a group of boys is beginning to be enforced. The fact that the boys are less burdened and so can range further afield in search of adventure, while the girls have to carry their heavy little charges with them, also makes a difference between the sexes. The groups of small children which hang about the fringes of some adult activity often contain both girls and boys, but here the association principle is simply age discrimination on the part of their elders, rather than voluntary association on the children's part.

These age gangs are usually confined to the children who

live in eight or ten contiguous households.* They are flexible chance associations, the members of which manifest a vivid hostility toward their contemporaries in neighbouring villages and sometimes toward other gangs within their own village. Blood ties cut across these neighbourhood alignments so that a child may be on good terms with members of two or three different groups. A strange child from another group, provided she came alone, could usually take refuge beside a relative. But the little girls of Siufaga looked askance at the little girls of Lumā, the nearest village and both looked with even greater suspicion at the little girls from Faleasao, who lived twenty minutes' walk away. However, heart burnings over these divisions were very temporary affairs. When Tua's brother was ill, her entire family moved from the far end of Siufaga into the heart of Lumā. For a few days Tua hung rather dolefully about the house, only to be taken in within a week by the central Lumā children with complete amiability. But when she returned some weeks later to Siufaga, she became again "a Siufaga girl," object elect of institutionalised scorn and gibes to her recent companions.

No very intense friendships are made at this age. The relationship and neighbourhood structure of the group overshadows the personalities within it. Also the most intense affection is always reserved for near relatives and pairs of little sisters take the place of chums. The Western comment, "Yes, Mary and Julia are such good friends as well as sisters!" becomes in Samoa, "But she is a relative," if a friendship is commented upon. The older ones fend for the younger, give them their spoils, weave them flower necklaces and give them their most treasured shells. This relationship aspect is the only permanent element in the group and even this is threatened by any change of residence. The emotional tone attached to the inhabitants of a strange village tends to make even a well-known cousin seem a little strange.

Of the different groups of little girls there was only one which showed characteristics which would make it possible to classify it as a gang. An accident of residence accounts for the most intense group development being in the centre of Lumā, where nine little girls of nearly the same age and with abundant relationship ties lived close together. The development of a group which played continually together and maintained a fairly coherent hostility towards outsiders, seems to be more of a function of residence than of the personality of any child particularly endowed with powers of leadership. The nine little girls in this group were less shy,

* See Neighbourhood Maps. Appendix I, page 147.

less suspicious, more generous towards one another, more socially enterprising than other children of the same age and in general reflected the socialising effects of group life. Outside this group, the children of this age had to rely much more upon their immediate relationship group reinforced perhaps by the addition of one or two neighbours. Where the personality of a child stood out it was more because of exceptional home environment than a result of social give-and-take with children of her own age.

Children of this age had no group activities except play, in direct antithesis to the home life where the child's only function was work—baby-tending and the performance of numerous trivial tasks and innumerable errands. They foregathered in the early evening, before the late Samoan supper, and occasionally during the general siesta hour in the afternoon. On moonlight nights they scoured the villages alternately attacking or fleeing from the gangs of small boys, peeking through drawn shutters, catching land crabs, ambushing wandering lovers, or sneaking up to watch a birth or a miscarriage in some distant house. Possessed by a fear of the chiefs, a fear of small boys, a fear of their relatives and by a fear of ghosts, no gang of less than four or five dared to venture forth on these nocturnal excursions. They were veritable groups of little outlaws escaping from the exactions of routine tasks. Because of the strong feeling for relationship and locality, the part played by stolen time, the need for immediately executed group plans, and the punishment which hung over the heads of children who got too far out of earshot, the Samoan child was as dependent upon the populousness of her immediate locality, as is the child in a rural community in the West. True her isolation here was never one-eighth of a mile in extent, but glaring sun and burning sands, coupled with the number of relatives to be escaped from in the day or the number of ghosts to be escaped from at night, magnified this distance until as a barrier to companionship it was equivalent to three or four miles in rural America. Thus there occurred the phenomenon of the isolated child in a village full of children of her own age. Such was Luna, aged ten, who lived in one of the scattered houses belonging to a high chief's household. This house was situated at the very end of the village where she lived with her grandmother, her mother's sister Sami, Sami's husband and baby, and two younger maternal aunts, aged seventeen and fifteen. Luna's mother was dead. Her other brothers and sisters lived on another island with her father's people. She was ten, but young for her age, a quiet, listless child, reluctant to take the initiative, the sort of child who would always need an institutionalised group life. Her only relatives close by were two girls of fourteen, whose long legs and absorption in semi-adult tasks made them far too grown-up companions for her. Some little girls of fourteen might have tolerated Luna about, but not Selu, the younger of the cousins, whose fine mat was already three feet under way. In the next house, a stone's throw away, lived two little girls, Pimi and Vana, aged eight and ten. But they were not relatives and being chief baby-tenders to four younger children, they had no time for exploring. There were no common relatives to bring them together and so Luna lived a solitary life, except when an enterprising young aunt of eleven came home to her mother's house. This aunt, Siva, was a fascinating companion, a vivid and precocious child, whom Luna followed about in open-mouthed astonishment. But Siva had proved too much of a handful for her widowed mother, and the *matai*, her uncle, had taken her to live in his immediate household at the other end of the village, on the other side of the central Lumā gang. They formed far more attractive companions and Siva seldom got as far as her mother's house in her occasional moments of freedom. So unenterprising Luna cared for her little cousin, followed her aunt and grandmother about and most of the time presented a very forlorn appearance.

In strong contrast was the fate of Lusi, who was only seven, too young to be really eligible for the games of her ten- and eleven-year-old elders. Had she lived in an isolated spot, she would have been merely a neighbourhood baby. But her house was in a strategic position, next door to that of her cousins, Maliu and Pola, important members of the Lumā gang. Maliu, one of the oldest members of the group, had a tremendous feeling for all her young relatives, and Lusi was her first cousin. So tiny, immature Lusi had the full benefit of a group life denied to Luna.

At the extreme end of Siufaga lived Vina, a gentle, unassuming girl of fourteen. Her father's house stood all alone in the centre of a grove of palm trees, just out of sight and ear-shot of the nearest neighbour. Her only companions were her first cousins of seventeen and nineteen. There was one little cousin of twelve also in the neighbourhood, but five younger brothers and sisters kept her busy. Vina also had several brothers and sisters younger than herself, but they were old enough to fend for themselves and Vina was comparatively free to follow the older girls on fishing expeditions. So she never escaped from being the little girl, tagging after older ones, carrying their loads and running their errands. She was a flurried anxious child, overconcerned with pleasing others, docile in her chance encounters with contemporaries from long habit of docility. A free give-and-take relationship within her own age group had been denied to her and was now denied to her forever. For it was only to the eight- to twelve-year-old girl that this casual group association was possible. As puberty approached, and a girl gained physical strength and added skill, her household absorbed her again. She must make the oven, she must go to work on the plantation, she must fish. Her days were filled with long tasks and new responsibilities.

Such a child was Fitu. In September she was one of the dominant members of the gang, a little taller than the rest, a little lankier, more strident and executive, but very much a harum-scarum little girl among little girls, with a great baby always on her hip. But by April she had turned the baby over to a younger sister of nine; the still younger baby was entrusted to a little sister of five and Fitu worked with her mother on the plantations, or went on long expeditions after hibiscus bark, or for fish. She took the family washing to the sea and helped make the oven on cooking days. Occasionally in the evening she slipped away to play games on the green with her former companions but usually she was too tired from the heavy unaccustomed work, and also a slight strangeness had crept over her. She felt that her more adult activities set her off from the rest of the group with whom she had felt so much at home the fall before. She made only abortive attempts to associate with the older girls in the neighbourhood. Her mother sent her to sleep in the pastor's house next door, but she returned home after three days. Those girls were all too old, she said. "Laititi a'u" ("I am but young"). And yet she was spoiled for her old group. The three villages numbered fourteen such children, just approaching puberty, preoccupied by unaccustomed tasks and renewed and closer association with the adults of their families, not yet interested in boys, and so forming no new alliances in accordance with sex interests. Soberly they perform their household tasks, select a teacher from the older women of their family, learn to bear the suffix, meaning "little," dropped from the "little girl" which had formerly described them. But they never again amalgamate into such free and easy groups as the before-the-teens gang. As sixteen- and seventeen-year-old girls, they will rely upon relatives, and the picture is groupings of twos or of threes, never more. The neighbourhood feelings drop out and girls of seventeen will ignore a near neighbour who is an age mate and go the length of the village to visit a relative. Relationship and similar sex interests are now the deciding factor in friendships. Girls also followed passively the stronger allegiance of the boys. If a girl's sweetheart has a chum who is interested in a cousin of hers, the girls strike up a lively, but temporary, friendship. Occasionally such friendships even go outside of the relationship group.

Although girls may confide only in one or two girl relatives, their sex status is usually sensed by the other women of the village and alliances shift and change on this basis, from the shy adolescent who is suspicious of all older girls, to the girl whose first or second love affair still looms as very important, to the girls who are beginning to centre all their attention upon one boy and possibly matrimony. Finally the unmarried mother selects her friends, when possible, from those in like case with herself, or from women of ambiguous marital position, deserted or discredited young wives.

Very few friendships of younger for older girls cut across these groupings after puberty. The twelve-year-old may have a great affection and admiration for her sixteen-year-old cousin (although any of these enthusiasms for older girls are pallid matters compared to a typical school girl "crush" in our civilisation). But when she is fifteen and her cousin nineteen, the picture changes. All of the adult and near-adult world is hostile, spying upon her love affairs in its more circumspect sophistication, supremely not to be trusted. No one is to be trusted who is not immediately engaged in similarly hazardous adventures.

It may safely be said that without the artificial conditions produced by residence in the native pastor's household or in the large missionary boarding school, the girls do not go outside their relationship group to make friends. (In addition to the large girls' boarding school which served all of American Samoa, the native pastor of each community maintained a small informal boarding school for boys and girls. To these schools were sent the girls whose fathers wished to send them later to the large boarding school, and also girls whose parents wished them to have three or four years of the superior educational advantages and stricter supervision of the pastor's home.) Here unrelated girls live side by side sometimes for years. But as one of the two defining features of a household is common residence, the friendships formed between girls who have lived in the pastor's household are not very different psychologically from the friendship of cousins or girls connected only by affinity who live in the same family. The only friendships which really are different in kind from those formed by common residence or membership in the same relationship group, are the institutionalised relationships between the wives of chiefs and the wives of talking chiefs. But these friendships can only be understood in connection with the friendships among boys and men.

The little boys follow the same pattern as do the little girls, running in a gang based upon the double bonds of neighbourhood and relationship. The feeling for the ascendency of age is always much stronger than in the case of girls because the older boys do not withdraw into their family groups as do the girls. The fifteen- and sixteen-year-old boys gang together with the same freedom as do the twelve-year-olds. The borderline between small boys and bigger boys is therefore a continually shifting one, the boys in an intermediate position now lording it over the younger boys, now tagging obsequiously in the wake of their elders. There are two institutionalised relationships between boys which are called by the same name and possibly were at one time one relationship. This is the *soa*, the companion at circumcision and the ambassador in love affairs. Boys are circumcised in pairs, making the arrangements themselves and seeking out an older man who has acquired a reputation for skilfulness. There seems to be here simply a logical inter-relationship of cause and effect; a boy chooses a friend (who is usually also a relative) as his companion and the experience shared binds them closer together. There were several pairs of boys in the village who had been circumcised together and were still inseparable companions, often sleeping together in the house of one of them. Casual homosexual practices occurred in such relationships. However, when the friendships of grown boys of the village were analysed, no close correspondence with the adolescent allegiance was found and older boys were as often found in groups of three or four as in pairs.

When a boy is two or three years past puberty, his choice of a companion is influenced by the convention that a young man seldom speaks for himself in love and never in a proposal of marriage. He accordingly needs a friend of about his age whom he can trust to sing his praises and press his suit with requisite fervour and discretion. For this office, a relative, or, if the affair be desperate, several relatives are employed. A youth is influenced in his choice by his need of an ambassador who is not only trustworthy and devoted but plausible and insinuating as a procurer. This *soa* relationship is often, but not necessarily, reciprocal. The expert in love comes in time to dispense with the services of an intermediary, wishing to taste to the full the sweets of all the stages in courtship. At the same time his services are much in demand by others, if they entertain any hope at all of his dealing honourably by his principal.

But the boys have also other matters besides love-making in which they must co-operate. Three are needed to man a bonito canoe; two usually go together to lasso eels on the reef; work on the communal taro plantations demands the labour of all the youths in the village. So that while a boy too chooses his best friends from among his relatives, his sense of social solidarity is always much stronger than that of a girl. The *Aualuma,* the organisation of young girls and wives of untitled men, is an exceedingly loose association gathered for very occasional communal work, and still more occasional festivities. In villages where the old intricacies of the social organisation are beginning to fall into disuse, it is the *Aualuma* which disappears first, while the *Aumaga,* the young men's organisation, has too important a place in the village economy to be thus ignored. The *Aumaga* is indeed the most enduring social factor in the village. The *matais* meet more formally and spend a great deal of time in their own households, but the young men work together during the day, feast before and after their labours, are present as a serving group at all meetings of the *matais,* and when the day's work is over, dance and go courting together in the evening. Many of the young men sleep in their friends' houses, a privilege but grudgingly accorded the more chaperoned girls.

Another factor which qualified men's relationships is the reciprocal relationship between chiefs and talking chiefs. The holders of these two classes of titles are not necessarily related, although this is often the case as it is considered an advantage to be related to both ranks. But the talking chiefs are major domos, assistants, ambassadors, henchmen, and councillors of their chiefs, and these relationships are often foreshadowed among the young men, the heirs-apparent or the heirs aspirant to the family titles.

Among women there are occasional close alliances between the *taupo* and the daughter of her father's principal talking chief. But these friendships always suffer from their temporary character; the *taupo* will inevitably marry into another village. And it is rather between the wife of the chief and the wife of a talking chief that the institutionalised and life-long friendship exists. The wife of the talking chief acts as assistant, advisor, and mouthpiece for the chief's wife and in turn counts upon the chief's wife for support and material help. It is a friendship based upon reciprocal obligations having their origins in the relationship between the women's husbands, and it is the only women's friendship which oversteps the limits of the relationship and affinity group. Such friendships based on an accident of marriage and enjoined by the social structure should hardly be classed as voluntary. And within the relationship group itself, friendship is so patterned as to be meaningless. I once asked a young married woman if a neighbour with whom she was always upon the most uncertain and irritated terms was a friend of hers. "Why, of course, her mother's father's father, and my father's mother's father were brothers." Friendship based on temperamental congeniality was a most tenuous bond, subject to shifts of residence, and a woman came to rely more and more on the associates to whose society and interest blood and marriage entitled her.

Association based upon age as a principle may be said to have ceased for the girls before puberty, due to the exceedingly individual nature of their tasks and the need for secrecy in their amatory adventures. In the case of the boys, greater freedom, a more compelling social structure, and continuous participation in co-operative tasks, brings about an age-group association which lasts through life. This grouping is influenced but not determined by relationship, and distorted by the influence of rank, prospective rank in the case of youth, equal rank but disproportionate age in the case of older men.

6

The Girl in the Community

THE COMMUNITY ignores both boys and girls from birth until they are fifteen or sixteen years of age. Children under this age have no social standing, no recognised group activities, no part in the social life except when they are conscripted for the informal dance floor. But at a year or two beyond puberty— the age varies from village to village so that boys of sixteen will in one place still be classed as small boys, in another as *taule'ale'as*, young men—both boys and girls are grouped into a rough approximation of the adult groupings, given a name for their organisation, and are invested with definite obligations and privileges in the community life.

The organisation of young men, the *Aumaga*, of young girls and the wives of untitled men and widows, the *Aualuma*, and of the wives of titled men, are all echoes of the central political structure of the village, the *Fono*, the organisation of *matais*, men who have the titles of chiefs or of talking chiefs. The *Fono* is always conceived as a round house in which each title has a special position, must be addressed with certain ceremonial phrases, and given a fixed place in the order of precedence in the serving of the kava. This ideal house has certain fixed divisions, in the right sector sit the high chief and his special assistant chiefs; in the front of the house sit the talking chiefs whose business it is to make the speeches, welcome strangers, accept gifts, preside over the distribution of food and make all plans and arrangements for group activities. Against the posts at the back of the house sit the *matais* of low rank, and between the posts and at the centre sit those of so little importance that no place is reserved for them. This framework of titles continues from generation to generation and holds a fixed place in the larger ideal structure of the titles of the whole island, the whole archipelago, the whole of Samoa. With some of these titles, which are in the gift of certain families, go certain privileges, a right to a house name, a right to confer a *taupo* name, a princess title, upon some young girl relative and an heir-apparent title, the *manaia*, on some boy of the household. Besides these prerogatives of the high chiefs, each member of the two classes of *matais*, chiefs and talking chiefs, has certain ceremonial rights. A talking chief must be served his kava with a special gesture, must be addressed with a separate set of verbs and nouns suitable to his rank, must be rewarded by the chiefs in tapa or fine mats for his ceremonially rendered services. The chiefs must be addressed with still another set of nouns and verbs, must be served with a different and more honourable gesture in the kava ceremony, must be furnished with food by their talking chiefs, must be honoured and escorted by the talking chiefs on every important occasion. The name of the village, the ceremonial name of the public square in which great ceremonies are held, the name of the meeting house of the *Fono*, the names of the principal chiefs and talking chiefs, the names of *taupo* and *manaia*, of the *Aualuma* and the *Aumaga*, are contained in a set of ceremonial salutations called the *Fa'alupega*, or courtesy titles of a village or district. Visitors on formally entering a village must recite the *Fa'alupega* as their initial courtesy to their hosts.

The *Aumaga* mirrors this organisation of the older men. Here the young men learn to make speeches, to conduct themselves with gravity and decorum, to serve and drink the kava, to plan and execute group enterprises. When a boy is old enough to enter the *Aumaga*, the head of his household either sends a present of food to the group, announcing the addition of the boy to their number, or takes him to a house where they are meeting and lays down a great kava root as a present. Henceforth the boy is a member of a group which is almost constantly together. Upon them falls all the heavy work of the village and also the greater part of the social intercourse between villages which centres about the young unmarried people. When a visiting village comes, it is the *Aumaga* which calls in a body upon the visiting *taupo*, taking gifts, dancing and singing for her benefit.

The organisation of the *Aualuma* is a less formalised version of the *Aumaga*. When a girl is of age, two or three years past puberty, varying with the village practice, her *matai* will send an offering of food to the house of the chief *taupo* of the village, thus announcing that he wishes the daughter of his house to be henceforth counted as one of the group of young girls who form her court. But while the *Aumaga* is centred about the *Fono*, the young men meeting outside or in a separate house, but exactly mirroring the forms and ceremonies of their elders, the *Aualuma* is centred about the person of the *taupo*, forming a group of maids of honour. They have no organisation as have the *Aumaga*, and furthermore, they do hardly any work. Occasionally the young girls may be called upon to sew thatch or gather paper mulberry; more occasionally they plant and cultivate a paper mulberry crop, but their main function is to be ceremonial helpers for the meetings of the wives of *matais*, and village hostesses in inter-village life. In many parts of Samoa the *Aualuma* has fallen entirely to pieces and is only remembered in the greeting words that fall from the lips of a stranger. But if the *Aumaga* should disappear, Samoan village life would have to be entirely reorganised, for upon the ceremonial and actual work of the young and untitled men the whole life of the village depends.

Although the wives of *matais* have no organisation recognised in the *Fa'alupega* (courtesy titles), their association is finer and more important than that of the *Aualuma*. The wives of titled men hold their own formal meetings, taking their status from their husbands, sitting at their husbands' posts and drinking their husbands' kava. The wife of the highest chief receives highest honour, the wife of the principal talking chief makes the most important speeches. The women are completely dependent upon their husbands for their status in this village group. Once a man has been given a title, he can never go back to the *Aumaga*. His title may be taken away from him when he is old, or if he is inefficient, but a lower title will be given him that he may sit and drink his kava with his former associates. But the widow or divorced wife of a *matai* must go back into the *Aualuma*, sit with the young girls outside the house, serve the food and run the errands, entering the women's *fono* only as a servant or an entertainer.

The women's *fonos* are of two sorts: *fonos* which precede or follow communal work, sewing the thatch for a guest house, bringing the coral rubble for its floor or weaving fine mats for the dowry of the *taupo*; and ceremonial *fonos* to welcome visitors from another village. Each of these meetings was designated by its purpose, as a *falelaga*, a weaving bee, or an *'aiga fiafia tama'ita'i*, ladies' feast. The women are only recognised socially by the women of a visiting village but the *taupo* and her court are the centre of the recognition of both men and women in the *malaga*, the travelling party. And these wives of high chiefs have to treat their own *taupo* with great courtesy and respect, address her as "your highness," accompany her on journeys, use a separate set of nouns and verbs when speaking to her. Here then is a discrepancy in which the young girls who are kept in strict subjection within their households, outrank their aunts and mothers in the social life between villages. This ceremonial undercutting of the older women's authority might seriously jeopardise the discipline of the household, if it were not for two considerations. The first is the tenuousness of the girls' organisation, the fact that within the village their chief *raison d'être* is to dance attendance upon the older women, who have definite industrial tasks to perform for the village; the second is the emphasis upon the idea of service as the chief duty of the *taupo*. The village princess is also the village servant. It is she who waits upon strangers, spreads their beds and makes their kava, dances when they wish it, and rises from her sleep to serve either the visitors or her own chief. And she is compelled to serve the social needs of the women as well as the men. Do they decide to borrow thatch in another village, they dress their *taupo* in her best and take her along to decorate the *malaga*. Her marriage is a village matter, planned and carried through by the talking chiefs and their wives who are her counsellors and chaperons. So that the rank of the *taupo* is really a further daily inroad upon her freedom as an individual, while the incessant chaperonage to which she is subjected and the way in which she is married without regard to her own wishes are a complete denial of her personality. And similarly, the slighter prestige of her untitled sisters, whose chief group activity is waiting upon their elders, has even less real significance in the daily life of the village.

With the exception of the *taupo,* the assumption of whose title is the occasion of a great festival and enormous distribution of property by her chief to the talking chiefs who must hereafter support and confirm her rank, a Samoan girl of good family has two ways of making her début. The first, the formal entry into the *Aualuma* is often neglected and is more a formal fee to the community than a recognition of the girl herself. The second way is to go upon a *malaga,* a formal travelling party. She may go as a near relative of the *taupo* in which case she will be caught up in a whirl of entertainment with which the young men of the host village surround their guests; or she may travel as the only girl in a small travelling party in which case she will be treated as a *taupo.* (All social occasions demand the presence of a *taupo,* a *manaia,* and a talking chief; and if individuals actually holding these titles are not present, some one else has to play the rôle.) Thus it is in inter-village life, either as a member of the *Aualuma* who call upon and dance for the *manaia* of the visiting *malaga*, or as a visiting girl in a strange village, that the unmarried Samoan girl is honoured and recognised by her community.

But these are exceptional occasions. A *malaga* may come only once a year, especially in Manu'a which numbers only seven villages in the whole archipelago. And in the daily life of the village, at crises, births, deaths, marriages, the unmarried girls have no ceremonial part to play. They are simply included with the "women of the household" whose duty it is to prepare the layette for the new baby, or carry stones to strew on the new grave. It is almost as if the community by its excessive recognition of the girl as a *taupo* or member of the *Aualuma*, considered itself exonerated from paying any more attention to her.

This attitude is fostered by the scarcity of taboos. In many parts of Polynesia, all women, and especially menstruating women, are considered contaminating and dangerous. A continuous rigorous social supervision is necessary, for a society can no more afford to ignore its most dangerous members than it can afford to neglect its most valuable. But in Samoa a girl's power of doing harm is very limited. She cannot make *tafolo*, a bread-fruit pudding usually made by the young men in any case, nor make the kava while she is menstruating. But she need retire to no special house; she need not eat alone; there is no contamination in her touch or look. In common with the young men and the older women, a girl gives a wide berth to a place where chiefs are engaged in formal work, unless she has special business there. It is not the presence of a woman which is interdicted but the uncalled-for intrusion of any one of either sex. No woman can be officially present at a gathering of chiefs unless she is *taupo* making the kava, but any woman may bring her husband his pipe or come to deliver a message, so long as her presence need not be recognised. The only place where a woman's femininity is in itself a real source of danger is in the matter of fishing canoes and fishing tackle which she is forbidden to touch upon pain of spoiling the fishing. But the enforcement of this prohibition is in the hands of individual fishermen in whose houses the fishing equipment is kept.

Within the relationship group matters are entirely different. Here women are very specifically recognised. The oldest female progenitor of the line, that is, the sister of the last holder of the title, or his predecessor's sister, has special rights over the distribution of the dowry which comes into the household. She holds the veto in the selling of land and other important family matters. Her curse is the most dreadful a man can incur for she has the power to "cut the line" and make the name extinct. If a man falls ill, it is his sister who must first take the formal oath that she has wished him no harm, as anger in her heart is most potent for evil. When a man dies, it is his paternal aunt or his sister who prepares the body for burial, anointing it with turmeric and rubbing it with oil, and it is she who sits beside the body, fanning away the flies, and keeps the fan in her possession ever after. And in the more ordinary affairs of the household, in the economic arrangements between relatives, in disputes over property or in family feuds, the women play as active a part as the men.

The girl and woman repays the general social negligence which she receives with a corresponding insouciance. She treats the lore of the village, the genealogies of the titles, the origin myths and local tales, the intricacies of the social organisation with supreme indifference. It is an exceptional girl who can give her great-grandfather's name, the exceptional boy who cannot give his genealogy in traditional form for several generations. While the boy of sixteen or seventeen is eagerly trying to master the esoteric allusiveness of the talking chief whose style he most admires, the girl of the same age learns the minimum of etiquette. Yet this is in no wise due to lack of ability. The *taupo* must have a meticulous knowledge, not only of the social arrangements of her own village, but also of those of neighbouring villages. She must serve visitors in proper form and with no hesitation after the talking chief has chanted their titles and the names of their kava cups. Should she take the wrong post which is the prerogative of another *taupo* who outranks her, her hair will be soundly pulled by her rival's female attendants. She learns the intricacies of the social organisation as well as her brother does. Still more notable is the case of the wife of a talking chief. Whether she is chosen for her docility by a man who has already assumed his title, or whether, as is often the case, she marries some boy of her acquaintance who later is made a talking chief, the *tausi*, wife of a talking chief, is quite equal to the occasion. In the meetings of women she must be a master of etiquette and the native rules of order, she must interlard her speeches with a wealth of unintelligible traditional material and rich allusiveness, she must preserve the same even voice, the same lofty demeanour, as her husband. And ultimately, the wife of an important talking chief must qualify as a teacher as well as a performer, for it is her duty to train the *taupo*. But unless the community thus recognises her existence, and makes formal demand upon her time and ability, a woman gives to it a bare minimum of her attention.

In like manner, women are not dealt with in the primitive penal code. A man who commits adultery with a chief's wife was beaten and banished, sometimes even drowned by the outraged community, but the woman was only cast out by her husband. The *taupo* who was found not to be a virgin was simply beaten by her female relatives. To-day if evil befalls the village, and it is attributed to some unconfessed sin on the part of a member of the community, the *Fono* and the *Aumaga* are convened and confession is enjoined upon any one who may have evil upon his conscience, but no such demand is made upon the *Aualuma* or the wives of the *matais*. This is in striking contrast to the family confessional where the sister is called upon first.

In matters of work the village makes a few precise demands. It is the women's work to cultivate the sugar cane and sew the thatch for the roof of the guest house, to weave the palm leaf blinds, and bring the coral rubble for the floor. When the girls have a paper mulberry plantation, the *Aumaga* occasionally help them in the work, the girls in turn making a feast for the boys, turning the whole affair into an industrious picnic. But between men's formal work and women's formal work there is a rigid division. Women do not enter into house-building or boat-building activities, nor go out in fishing canoes, nor may men enter the formal weaving house or the house where women are making tapa in a group. If the women's work makes it necessary for them to cross the village, as is the case when rubble is brought up from the seashore to make the floor of the guest house, the men entirely disappear, either gathering in some remote house, or going away to the bush or to another village. But this avoidance is only for large formal occasions. If her husband is building the family a new cook-house, a woman may make tapa two feet away, while a chief may sit and placidly braid cinet while his wife weaves a fine mat at his elbow.

So, although unlike her husband and brothers a woman spends most of her time within the narrower circle of her household and her relationship group, when she does participate in community affairs she is treated with the punctilio which marks all phases of Samoan social life. The better part of her attention and interest is focused on a smaller group, cast in a more personal mode. For this reason, it is impossible to evaluate accurately the difference in innate social drive between men and women in Samoa. In those social spheres where women have been given an opportunity, they take their place with as much ability as the men. The wives of the talking chiefs in fact exhibit even greater adaptability than their husbands. The talking chiefs are especially chosen for their oratorical and intellectual abilities, whereas the women have a task thrust upon them at their marriage requiring great oratorical skill, a fertile imagination, tact, and a facile memory.

7

Formal Sex Relations

THE FIRST ATTITUDE which a little girl learns towards boys is one of avoidance and antagonism. She learns to observe the brother and sister taboo towards the boys of her relationship group and household, and together with the other small girls of her age group she treats all other small boys as enemies elect. After a little girl is eight or nine years of age she has learned never to approach a group of older boys. This feeling of antagonism towards younger boys and shamed avoidance of older ones continues up to the age of thirteen or fourteen, to the group of girls who are just reaching puberty and the group of boys who have just been circumcised. These children are growing away from the age-group life and the age-group antagonisms. They are not yet actively sex-conscious. And it is at this time that relationships between the sexes are least emotionally charged. Not until she is an old married woman with several children will the Samoan girl again regard the opposite sex so quietly. When these adolescent children gather together there is a good-natured banter, a minimum of embarrassment, a great deal of random teasing which usually takes the form of accusing some little girl of a consuming passion for a decrepit old man of eighty, or some small boy of being the father of a buxom matron's eighth child. Occasionally the banter takes the form of attributing affection between two age mates and is gaily and indignantly repudiated by both. Children at this age meet at informal *siva* parties, on the outskirts of more formal occasions, at community reef fishings (when many yards of reef have been enclosed to make a great fish trap) and on torch-fishing excursions. Good-natured tussling and banter and co-operation in common activities are the keynotes of these occasions. But unfortunately these contacts are neither frequent nor sufficiently prolonged to teach the girls co-operation or to give either boys or girls any real appreciation of personality in members of the opposite sex.

Two or three years later this will all be changed. The fact that little girls no longer belong to age groups makes the individual's defection less noticeable. The boy who begins to take an active interest in girls is also seen less in a gang and spends more time with one close companion. Girls have lost all of their nonchalance. They giggle, blush, bridle, run away. Boys become shy, embarrassed, taciturn, and avoid the society of girls in the daytime and on the brilliant moonlit nights for which they accuse the girls of having an exhibitionistic preference. Friendships fall more strictly within the relationship group. The boy's need for a trusted confidant is stronger than that of the girl, for only the most adroit and hardened Don Juans do their own courting. There are occasions, of course, when two youngsters just past adolescence, fearful of ridicule, even from their nearest friends and relatives, will slip away alone into the bush. More frequently still an older man, a widower or a divorced man, will be a girl's first lover. And here there is no need for an ambassador. The older man is neither shy nor frightened, and furthermore there is no one whom he can trust as an intermediary; a younger man would betray him, an older man would not take his amours seriously. But the first spontaneous experiment of adolescent children and the amorous excursions of the older men among the young girls of the village are variants on the edge of the recognised types of relationships; so also is the first experience of a young boy with an older woman. But both of these are exceedingly frequent occurrences, so that the success of an amatory experience is seldom jeopardised by double ignorance. Nevertheless, all of these occasions are outside the recognised forms into which sex relations fall. The little boy and girl are branded by their companions as guilty of *tautala lai titi* (presuming above their ages) as is the boy who loves or aspires to love an older woman, while the idea of an older man pursuing a young girl appeals strongly to their sense of humour; or if the girl is very young and naïve, to their sense of unfitness. "She is too young, too young yet. He is too old," they will say, and the whole weight of vigorous disapproval fell upon a *matai* who was known to be the father of the child of Lotu, the sixteen-year-old feeble-minded girl on Olesega. Discrepancy in age or experience always strikes them as comic or pathetic according to the degree. The theoretical punishment which is meted out to a disobedient and runaway daughter is to marry her to a very old man, and I have heard a nine-year-old giggle contemptuously over her mother's preference for a seventeen-year-old boy. Worst among these unpatterned deviations is that of the man who makes love to some young and dependent woman of his household, his adopted child or his wife's younger sister. The cry of incest is raised against him and sometimes feeling runs so high that he has to leave the group.

Besides formal marriage there are only two types of sex relations which receive any formal recognition from the community—love affairs between unmarried young people (this includes the widowed) who are very nearly of the same age, whether leading to marriage or merely a passing diversion; and adultery.

Between the unmarried there are three forms of relationship: the clandestine encounter, "under the palm trees," the published elopement, *Avaga,* and the ceremonious courtship in which the boy "sits before the girl"; and on the edge of these, the curious form of surreptitious rape, called *moetotolo,* sleep crawling, resorted to by youths who find favour in no maiden's eyes.

In these three relationships, the boy requires a confidant and ambassador whom he calls a *soa.* Where boys are close companions, this relationship may extend over many love affairs, or it may be a temporary one, terminating with the particular love affair. The *soa* follows the pattern of the talking chief who makes material demands upon his chief in return for the immaterial services which he renders him. If marriage results from his ambassadorship, he receives a specially fine present from the bridegroom. The choice of a *soa* presents many difficulties. If the lover chooses a steady, reliable boy, some slightly younger relative devoted to his interests, a boy unambitious in affairs of the heart, very likely the ambassador will bungle the whole affair through inexperience and lack of tact. But if he chooses a handsome and expert wooer who knows just how "to speak softly and walk gently," then as likely as not the girl will prefer the second to the principal. This difficulty is occasionally anticipated by employing two or three *soas* and setting them to spy on each other. But such a lack of trust is likely to inspire a similar attitude in the agents, and as one overcautious and disappointed lover told me ruefully, "I had five *soas,* one was true and four were false."

Among possible *soas* there are two preferences, a brother or a girl. A brother is by definition loyal, while a girl is far more skilful for "a boy can only approach a girl in the evening, or when no one is by, but a girl can go with her all day long, walk with her and lie on the mat by her, eat off the same platter, and whisper between mouthfuls the name of the boy, speaking ever of him, how good he is, how gentle and how true, how worthy of love. Yes, best of all is the *soafafine,* the woman ambassador." But the difficulties of obtaining a *soafafine* are great. A boy may not choose from his own female relatives. The taboo forbids him ever to mention such matters in their presence. It is only by good chance that his brother's sweetheart may be a relative of the girl upon whom he has set his heart; or some other piece of good fortune may throw him into contact with a girl or woman who will act in his interests. The most violent antagonisms in the young people's groups are not between ex-lovers, arise not from the venom of the deserted nor the smarting pride of the jilted, but occur between the boy and the *soa* who has betrayed him, or a lover and the friend of his beloved who has in any way blocked his suit.

In the strictly clandestine love affair the lover never presents himself at the house of his beloved. His *soa* may go there in a group or upon some trumped-up errand, or he also may avoid the house and find opportunities to speak to the girl while she is fishing or going to and from the plantation. It is his task to sing his friend's praise, counteract the girl's fears and objections, and finally appoint a rendezvous. These affairs are usually of short duration and both boy and girl may be carrying on several at once. One of the recognised causes of a quarrel is the resentment of the first lover against his successor of the same night, "for the boy who came later will mock him." These clandestine lovers make their rendezvous on the outskirts of the village. "Under the palm trees" is the conventionalised designation of this type of intrigue. Very often three or four couples will have a common rendezvous, when either the boys or the girls are relatives who are friends. Should the girl ever grow faint or dizzy, it is the boy's part to climb the nearest palm and fetch down a fresh cocoanut to pour on her face in lieu of *eau de cologne*. In native theory, barrenness is the punishment of promiscuity; and, *vice versa*, only persistent monogamy is rewarded by conception. When a pair of clandestine experimenters whose rank is so low that their marriages are not of any great economic importance become genuinely attached to each other and maintain the relationship over several months, marriage often follows. And native sophistication distinguishes between the adept lover whose adventures are many and of short duration and the less skilled man who can find no better proof of his virility than a long affair ending in conception.

Often the girl is afraid to venture out into the night, infested with ghosts and devils, ghosts that strangle one, ghosts from far-away villages who come in canoes to kidnap the girls of the village, ghosts who leap upon the back and may not be shaken off. Or she may feel that it is wiser to remain at home, and if necessary, attest her presence vocally. In this case the lover braves the house; taking off his *lavalava*, he greases his body thoroughly with cocoanut oil so that he can slip through the fingers of pursuers and leave no trace, and stealthily raises the blinds and slips into the house. The prevalence of this practice gives point to the familiar incident in Polnesian folk tales of the ill fortune that falls the luckless hero who "sleeps until morning, until the rising sun reveals his presence to the other inmates of the house." As perhaps a dozen or more people and several dogs are sleeping in the house, a due regard for silence is sufficient precaution. But it is this habit of domestic rendez-vous which lends itself to the peculiar abuse of the *moetotolo*, or sleep crawler.

The *moetotolo* is the only sex activity which presents a definitely abnormal picture. Ever since the first contact with white civilisation, rape, in the form of violent assault, has occurred occasionally in Samoa. It is far less congenial, however, to the Samoan attitude than *moetotolo*, in which a man stealthily appropriates the favours which are meant for another. The need for guarding against discovery makes conversation impossible, and the sleep crawler relies upon the girl's expecting a lover or the chance that she will indiscriminately accept any comer. If the girl suspects and resents him, she raises a great outcry and the whole household gives chase. Catching a *moetotolo* is counted great sport, and the women, who feel their safety endangered, are even more active in pursuit than the men. One luckless youth in Luma neglected to remove his *lavalava*. The girl discovered him and her sister succeeded in biting a piece out of his *lavalava* before he escaped. This she proudly exhibited the next day. As the boy had been too dull to destroy his *lavalava*, the evidence against him was circumstantial and he was the laughing stock of the village; the children wrote a dance song about it and sang it after him wherever he went. The *moetotolo* problem is complicated by the possibility that a boy of the household may be the offender and may take refuge in the hue and cry following the discovery. It also provides the girl with an excellent alibi, since she has only to call out *"moetotolo"* in case her lover is discovered. "To the family and the village that may be a *moetotolo*, but it is not so in the hearts of the girl and the boy."

Two motives are given for this unsavoury activity, anger and failure in love. The Samoan girl who plays the coquette does so at her peril. "She will say, 'Yes, I will meet you to-night by that old cocoanut tree just besides the devilfish stone when the moon goes down.' And the boy will wait and wait and wait all night long. It will grow very dark; lizards will drop on his head; the ghost boats will come into the channel. He will be very much afraid. But he will wait there until dawn, until his hair is wet with dew and his heart is very angry and still she does not come. Then in revenge he will attempt a *moetotolo*. Especially will he do so if he hears that she has met another that night." The other set explanation is that a particular boy cannot win a sweetheart by any legitimate means, and there is no form of prostitution, except guest prostitution in Samoa. As some of the boys who were notorious *moetotolos* were among the most charming and good-looking youths of the village, this is a little hard to understand. Apparently, these youths, frowned upon in one or two tentative courtships, inflamed by the loudly proclaimed success of their fellows and the taunts against their own inexperience, cast established wooing procedure to the winds and attempt a *moetotolo*. And once caught, once branded, no girl will ever pay any attention to them again. They must wait until as older men, with position and title to offer, they can choose between some weary and bedraggled wanton or the unwilling young daughter of ambitious and selfish parents. But years will intervene before this is possible, and shut out from the amours in which his companions are engaging, a boy makes one attempt after another, sometimes successfully, sometimes only to be caught and beaten, mocked by the village, and always digging the pit deeper under his feet. Often partially satisfactory solutions are relationships with men. There was one such pair in the village, a notorious *moetotolo,* and a serious-minded youth who wished to keep his heart free for political intrigue. The *moetotolo* therefore complicates and adds zest to the surreptitious love-making which is conducted at home, while the danger of being missed, the undesirability of chance encounters abroad, rain and the fear of ghosts, complicate "love under the palm trees."

Between these strictly *sub rosa* affairs and a final offer of marriage there is an intermediate form of courtship in which the girl is called upon by the boy. As this is regarded as a tentative move towards matrimony, both relationship groups must be more or less favourably inclined towards the union. With his *soa* at his side and provided with a basket of fish, an octopus or so, or a chicken, the suitor presents himself at the girl's home before the late evening meal. If his gift is accepted, it is a sign that the family of the girl are willing for him to pay his addresses to her. He is formally welcomed by the *matai*, sits with reverently bowed head throughout the evening prayer, and then he and his *soa* stay for supper. But the suitor does not approach his beloved. They say: "If you wish to know who is really the lover, look then not at the boy who sits by her side, looks boldly into her eyes and twists the flowers in her necklace around his fingers or steals the hibiscus flower from her hair that he may wear it behind his ear. Do not think it is he who whispers softly in her ear, or says to her, 'Sweetheart, wait for me to-night. After the moon has set, I will come to you,' or who teases her by saying she has many lovers. Look instead at the boy who sits far off, who sits with bent head and takes no part in the joking. And you will see that his eyes are always turned softly on the girl. Always he watches her and never does he miss a movement of her lips. Perhaps she will wink at him, perhaps she will raise her eyebrows, perhaps she will make a sign with her hand. He must always be wakeful and watching or he will miss it." The *soa* meanwhile pays the girl elaborate and ostentatious court and in undertones pleads the cause of his friend. After dinner, the centre of the house is accorded the young people to play cards, sing or merely sit about, exchanging a series of broad pleasantries. This type of courtship varies from occasional calls to daily attendance. The food gift need not accompany each visit, but is as essential at the initial call as is an introduction in the West. The way of such declared lovers is hard. The girl does not wish to marry, nor to curtail her amours in deference to a definite betrothal. Possibly she may also dislike her suitor, while he in turn may be the victim of family ambition. Now that the whole village knows him for her suitor, the girl gratifies her vanity by avoidance, by perverseness. He comes in the evening, she has gone to another house; he follows her there, she immediately returns home. When such courtship ripens into an accepted proposal of marriage, the boy often goes to sleep in the house of his intended bride and often the union is surreptitiously consummated. Ceremonial marriage is deferred until such time as the boy's family have planted or collected enough food and other property and the girl's family have gotten together a suitable dowry of tapa and mats.

In such manner are conducted the love affairs of the average young people of the same village, and of the plebeian young people of neighbouring villages. From this free and easy experimentation, the *taupo* is excepted. Virginity is a legal requirement for her. At her marriage, in front of all the people, in a house brilliantly lit, the talking chief of the bridegroom will take the tokens of her virginity.* In former days should she prove not to be a virgin, her female relatives fell upon and beat her with stones, disfiguring and sometimes fatally injuring the girl who had shamed their house. The public ordeal sometimes prostrated the girl for as much as a week, although ordinarily a girl recovers from first intercourse in two or three hours, and women seldom lie abed more than a few hours after childbirth. Although this virginity-testing ceremony was theoretically observed at weddings of people of all ranks, it was simply ignored if the boy knew that it was an idle form, and "a wise girl who is not a virgin will tell the talking chief

---

* This custom is now forbidden by law, but is only gradually dying out.

of her husband, so that she be not shamed before all the people."

The attitude towards virginity is a curious one. Christianity has, of course, introduced a moral premium on chastity. The Samoans regard this attitude with reverent but complete scepticism and the concept of celibacy is absolutely meaningless to them. But virginity definitely adds to a girl's attractiveness, the wooing of a virgin is considered far more of a feat than the conquest of a more experienced heart, and a really successful Don Juan turns most of his attention to their seduction. One youth who at twenty-four married a girl who was still a virgin was the laughing stock of the village over his freely related trepidation which revealed the fact that at twenty-four, although he had had many love affairs, he had never before won the favours of a virgin.

The bridegroom, his relatives and the bride and her relatives all receive prestige if she proves to be a virgin, so that the girl of rank who might wish to forestall this painful public ceremony is thwarted not only by the anxious chaperonage of her relatives but by the boy's eagerness for prestige. One young Lothario eloped to his father's house with a girl of a high rank from another village and refused to live with her because, said he, "I thought maybe I would marry that girl and there would be a big *malaga* and a big ceremony and I would wait and get credit for marrying a virgin. But the next day her father came and said that she could not marry me, and she cried very much. So I said to her, 'Well, there is no use now to wait any longer. Now we will run away into the bush.' " It is conceivable that the girl would often trade the temporary prestige for an escape from the public ordeal, but in proportion as his ambitions were honourable, the boy would frustrate her efforts.

Just as the clandestine and casual "love under the palm trees" is the pattern irregularity for those of humble birth, so the elopement has its archetype in the love affairs of the *taupo,* and the other chiefs' daughters. These girls of noble birth are carefully guarded; not for them are secret trysts at night or stolen meetings in the day time. Where parents of lower rank complacently ignore their daughters' experiments, the high chief guards his daughter's virginity as he guards the honour of his name, his precedence in the kava ceremony or any other prerogative of his high degree. Some old woman of the household is told off to be the girl's constant companion and duenna. The *taupo* may not visit in other houses in the village, or leave the house alone at night. When she sleeps, an older woman sleeps by her side. Never may she go to another village unchaperoned. In her own village she goes soberly about her tasks, bathing in the sea, working in the plantation, safe under the jealous guardianship of the women of her own village. She runs little risk from the *moetotolo,* for one who outraged the *taupo* of his village would formerly have been beaten to death, and now would have to flee from the village. The prestige of the village is inextricably bound up with the high repute of the *taupo* and few young men in the village would dare to be her lovers. Marriage to them is out of the question, and their companions would revile them as traitors rather than envy them such doubtful distinction. Occasionally a youth of very high rank in the same village will risk an elopement, but even this is a rare occurrence. For tradition says that the *taupo* must marry outside her village, marry a high chief or a *manaia* of another village. Such a marriage is an occasion for great festivities and solemn ceremony. The chief and all of his talking chiefs must come to propose for her hand, come in person bringing gifts for her talking chiefs. If the talking chiefs of the girl are satisfied that this is a lucrative and desirable match, and the family are satisfied with the rank and appearance of the suitor, the marriage is agreed upon. Little attention is paid to the opinion of the girl. So fixed is the idea that the marriage of the *taupo* is the affair of the talking chiefs that Europeanised natives on the main island, refuse to make their daughters *taupos* because the missionaries say a girl should make her own choice, and once she is a *taupo,* they regard the matter as inevitably taken out of their hands. After the betrothal is agreed upon the bridegroom returns to his village to collect food and property for the wedding. His village sets aside a piece of land which is called the "Place of the Lady" and is her property and the property of her children forever, and on this land they build a house for the bride. Meanwhile, the bridegroom has left behind him in the house of the bride a talking chief, the counterpart of the humbler *soa*. This is one of the talking chief's best opportunities to acquire wealth. He stays as the emissary of his chief, to watch over his future bride. He works for the bride's family and each week the *matai* of the bride must reward him with a handsome present. As an affianced wife of a chief, more and more circumspect conduct is enjoined upon the girl. Did she formerly joke with the boys of the village, she must joke no longer, or the talking chief, on the watch for any lapse from high decorum, will go home to his chief and report that his bride is unworthy of such honour. This custom is particularly susceptible to second thought on the part of either side. Does the bridegroom repent of the bargain, he bribes his talking chief (who is usually a young man, not one of the important talking chiefs who will benefit greatly by the marriage itself) to be oversensitive to the behaviour of the bride or the treatment he receives in the bride's family. And this is the time in which the bride will elope, if her affianced husband is too unacceptable. For while no boy of her own village will risk her dangerous favours, a boy from another village will enormously enhance his prestige if he elopes with the *taupo* of a rival community. Once she has eloped, the projected alliance is of course broken off, although her angry parents may refuse to sanction her marriage with her lover and marry her for punishment to some old man.

So great is the prestige won by the village, one of whose young men succeeds in eloping with a *taupo,* that often the whole effort of a *malaga* is concentrated upon abducting the *taupo,* whose virginity will be respected in direct ratio to the chances of her family and village consenting to ratify the marriage. As the abductor is often of high rank, the village often ruefully accepts the compromise.

This elopement pattern, given meaning by the restrictions under which the *taupo* lives and this inter-village rivalry, is carried down to the lower ranks where indeed it is practically meaningless. Seldom is the chaperonage exercised over the girl of average family severe enough to make elopement the only way of consummating a love affair. But the elopement is spectacular; the boy wishes to increase his reputation as a successful Don Juan, and the girl wishes to proclaim her conquest and also often hopes that the elopement will end in marriage. The eloping pair run away to the parents of the boy or to some of his relatives and wait for the girl's relatives to pursue her. As one boy related the tale of such an adventure: "We ran away in the rain, nine miles to Leone, in the pouring rain, to my father's house. The next day her family came to get her, and my father said to me, 'How is it, do you wish to marry this girl, shall I ask her father to leave her here?' And I said, 'Oh, no. I just eloped with her for public information.'" Elopements are much less frequent than the clandestine love affairs because the girl takes far more risk. She publicly renounces her often nominal claims to virginity; she embroils herself with her family, who in former times, and occasionally even to-day, would beat her soundly and shave off her hair. Nine times out of ten, her lover's only motive is vanity and display, for the boys say, "The girls hate a *moetotolo,* but they all love an *avaga* (eloping) man."

The elopement also occurs as a practical measure when one family is opposed to a marriage upon which a pair of young people have determined. The young people take refuge with the friendly side of the family. But unless the recalcitrant family softens and consents to legalise the marriage by a formal exchange of property, the principals can do nothing to establish their status. A young couple may have had several children and still be classed as "elopers," and if the marriage is finally legalised after long delay, this stigma will always cling to them. It is far more serious a one than a mere accusation of sexual irregularity, for there is a definite feeling that the whole community procedure has been outraged by a pair of young upstarts.

Reciprocal gift-giving relations are maintained between the two families as long as the marriage lasts, and even afterwards if there are children. The birth of each child, the death of a member of either household, a visit of the wife to her family, or if he lives with her people, of the husband to his, is marked by the presentation of gifts.

In premarital relationships, a convention of love making is strictly adhered to. True, this is a convention of speech, rather than of action. A boy declares that he will die if a girl refuses him her favours, but the Samoans laugh at stories of romantic love, scoff at fidelity to a long absent wife or mistress, believe explicitly that one love will quickly cure another. The fidelity which is followed by pregnancy is taken as proof positive of a real attachment, although having many mistresses is never out of harmony with a declaration of affection for each. The composition of ardent love songs, the fashioning of long and flowery love letters, the invocation of the moon, the stars and the sea in verbal courtship, all serve to give Samoan love-making a close superficial resemblance to our own, yet the attitude is far closer to that of Schnitzler's hero in *The Affairs of Anatol*. Romantic love as it occurs in our civilisation, inextricably bound up with ideas of monogamy, exclusiveness, jealousy and undeviating fidelity does not occur in Samoa. Our attitude is a compound, the final result of many converging lines of development in Western civilisation, of the institution of monogamy, of the ideas of the age of chivalry, of the ethics of Christianity. Even a passionate attachment to one person which lasts for a long period and persists in the face of discouragement but does not bar out other relationships, is rare among the Samoans. Marriage, on the other hand, is regarded as a social and economic arrangement, in which relative wealth, rank, and skill of husband and wife, all must be taken into consideration. There are many marriages in which both individuals, especially if they are over thirty, are completely faithful. But this must be attributed to the ease of sexual adjustment on the one hand, and to the ascendency of other interests, social organisation for the men, children for the women, over sex interests, rather than to a passionate fixation upon the partner in the marriage. As the Samoans lack the inhibitions and the intricate specialisation of sex feeling which make marriages of convenience unsatisfactory, it is possible to bulwark marital happiness with other props than temporary passionate devotion. Suitability and expediency become the deciding factors.

Adultery does not necessarily mean a broken marriage. A chief's wife who commits adultery is deemed to have dishonoured her high position, and is usually discarded, although the chief will openly resent her remarriage to any one of lower rank. If the lover is considered the more culpable, the village will take public vengeance upon him. In less conspicuous cases the amount of fuss which is made over adultery is dependent upon the relative rank of the offender and offended, or the personal jealousy which is only occasionally aroused. If either the injured husband or the injured wife is sufficiently incensed to threaten physical violence, the trespasser may have to resort to a pubic *ifoga*, the ceremonial humiliation before some one whose pardon is asked. He goes to the house of the man he has injured, accompanied by all the men of his household, each one wrapped in a fine mat, the currency of the country; the suppliants seat themselves outside the house, fine mats spread over their heads, hands folded on their breasts, heads bent in attitudes of the deepest dejection and humiliation. "And if the man is very angry he will say no word. All day he will go about his business; he will braid cinet with a quick hand, he will talk loudly to his wife, and call out greetings to those who pass in the roadway, but will take no notice of those who sit on his own terrace, who dare not raise their eyes or make any movement to go away. In olden days, if his heart was not softened, he might take a club and together with his relatives go out and kill those who sit without. But now he only keeps them waiting, waiting all day long. The sun will beat down upon them; the rain will come and beat on their heads and still he will say no word. Then towards evening he will say at last: 'Come, it is enough. Enter the house and drink the kava. Eat the food which I will set before you and we will cast our trouble into the sea.' " Then the fine mats are accepted as payment for the injury, the *ifoga* becomes a matter of village history and old gossips will say, "Oh, yes, Lua! no, she's not Iona's child. Her father is that chief over in the next village. He *ifod* to Iona before she was born." If the offender is of much lower rank than the injured husband, his chief, or his father (if he is only a young boy) will have to humiliate himself in his place. Where the offender is a woman, she and her female relatives will make similar amends. But they will run far greater danger of being roundly beaten and berated; the peaceful teachings of Christianity—perhaps because they were directed against actual killing, rather than the slightly less fatal encounters of women—have made far less change in the belligerent activities of the women than in those of the men.

If, on the other hand, a wife really tires of her husband, or a husband of his wife, divorce is a simple and informal matter, the non-resident simply going home to his or her family, and the relationship is said to have "passed away." It is a very brittle monogamy, often trespassed and more often broken entirely. But many adulteries occur—between a young marriage-shy bachelor and a married woman, or a temporary widower and some young girl—which hardly threaten the continuity of established relationships. The claim that a woman has on her family's land renders her as independent as her husband, and so there are no marriages of any duration in which either person is actively unhappy. A tiny flare-up and a woman goes home to her own people; if her husband does not care to conciliate her, each seeks another mate.

Within the family, the wife obeys and serves her husband, in theory, though, of course, the hen-pecked husband is a frequent phenomenon. In families of high rank, her personal service to her husband is taken over by the *taupo* and the talking chief but the wife always retains the right to render a high chief sacred personal services, such as cutting his hair. A wife's rank can never exceed her husband's because it is always directly dependent upon it. Her family may be richer and more illustrious than his, and she may actually exercise more influence over the village affairs through her blood relatives than he, but within the life of the household and the village, she is a *tausi*, wife of talking chief, or a *faletua*, wife of a chief. This sometimes results in conflict, as in the case of Pusa who was the sister of the last holder of the highest title on the island. This title was temporarily extinct. She was also the wife of the highest chief in the village. Should her brother, the heir, resume the higher title, her husband's rank and her rank as his wife would suffer. Helping her brother meant lowering the prestige of her husband. As she was the type of woman who cared a great deal more for wire pulling than for public recognition, she threw her influence in for her brother. Such conflicts are not uncommon, but they present a clear-cut choice, usually reinforced by considerations of residence. If a woman lives in her husband's household, and if, furthermore, that household is in another village, her interest is mainly enlisted in her husband's cause; but if she lives with her own family, in her own village, her allegiance is likely to cling to blood relatives from whom she receives reflected glory and informal privilege, although no status.

8

The Rôle of the Dance

DANCING IS the only activity in which almost all ages and both sexes participate and it therefore offers a unique opportunity for an analysis of education.

In the dance there are virtuosos but no formal teachers. It is a highly individual activity set in a social framework. This framework varies from a small dancing party at which twelve to twenty people are present to the major festivities of a *malaga* (travelling party) or a wedding when the largest guest house in the village is crowded within and encircled by spectators without. With the size and importance of the festivity, the formality of the arrangements varies also. Usually the occasion of even a small *siva* (dance) is the presence of at least two or three strange young people from another village. The pattern entertainment is a division of the performers into visitors and hosts, the two sides taking turns in providing the music and dancing. This pattern is still followed even when the *malaga* numbers only two individuals, a number of hosts going over to swell the visitors' ranks.

It is at these small informal dances that the children learn to dance. In the front of the house sit the young people who are the centre and arbiters of the occasion. The *matai* and his wife and possibly a related *matai* and the other elders of the household sit at the back of the house, in direct reversal of the customary procedure according to which the place of the young people is in the background. Around the ends cluster women and children, and outside lurk the boys and girls who are not participating in the dancing, although at any moment they may be drawn into it. On such occasions the dancing is usually started by the small children, beginning possibly with seven- and eight-year olds. The chief's wife or one of the young men will call out the names of the children and they are stood up in a group of three, sometimes all boys or girls, sometimes with a girl between two boys, which is the conventional adult grouping for the *taupo* and her two talking chiefs. The young men, sitting in a group near the centre of the house, provide the music, one of them standing and leading the singing to the accompaniment of an imported stringed instrument which has taken the place of the rude bamboo drum of earlier times. The leader sets the key and the whole company join in either in the song, or by clapping, or by beating on the floor with their knuckles. The dancers themselves are the final arbiters of the excellence of the music and it is not counted as petulance for a dancer to stop in the middle and demand better music as the price of continuing. The songs sung are few in number; the young people of one village seldom know more than a dozen airs; and perhaps twice as many sets of words which are sung now to one air, now to another. The verse pattern is simply based upon the number of syllables; a change in stress is permitted and rhyme is not demanded so that any new event is easily set in the old pattern, and names of villages and of individuals are inserted with great freedom. The content of the songs is likely to take on an extremely personal character containing many quips at the expense of individuals and their villages.

The form of the participation of the audience changes according to the age of the dancers. In the case of the smaller children, it consists of an endless stream of good-natured comment: "Faster!" "Down lower! Lower!" "Do it again!" "Fasten your *lavalava*." In the dancing of the more expert boys and girls the group takes part by a steady murmur of "Thank you, thank you, for your dancing!" "Beautiful! Engaging! Charming! Bravo!" which gives very much the effect of the irregular stream of "Amens" at an evangelistic revival. This articulate courtesy becomes almost lyric in quality when the dancer is a person of rank for whom dancing at all is a condescension.

The little children are put out upon these public floors with a minimum of preliminary instruction. As babies in their mothers' arms at just such a party as this, they learned to clap before they learned to walk, so that the beat is indelibly fixed in their minds. As two- and three-year-olds they have stood on a mat at home and clapped their hands in time to their elders' singing. Now they are called upon to perform before a group. Wide-eyed, terrified babies stand beside some slightly older child, clapping in desperation and trying to add new steps borrowed on the spur of the moment from their companions. Every improvement is greeted with loud applause. The child who performed best at the last party is haled forward at the next, for the group is primarily interested in its own amusement rather than in distributing an equal amount of practice among the children. Hence some children rapidly outdistance the rest, through interest and increased opportunity as well as superior gift. This tendency to give the talented child another and another chance is offset somewhat by rivalry between relatives who wish to thrust their little ones forward.

While the children are dancing, the older boys and girls are refurbishing their costumes with flowers, shell necklaces, anklets and bracelets of leaves. One or two will probably slip off home and returned dressed in elaborate bark skirts. A bottle of cocoanut oil is produced from the family chest and rubbed on the bodies of the older dancers. Should a person of rank be present and consent to dance, the hostess family bring out their finest mats and tapas as costume. Sometimes this impromptu dressing assumes such importance that an adjoining house is taken over as a dressing room; at others it is of so informal a nature that spectators, who have gathered outside arrayed only in sheets, have to borrow a dress or a *lavalava* from some other spectator before they can appear on the dance floor.

The form of the dance itself is eminently individualistic. No figures are prescribed except the half dozen formal little claps which open the dance and the use of one of a few set endings. There are twenty-five or thirty figures, two or three set transitional positions, and at least three definite styles, the dance of the *taupo*, the dance of the boys, and the dance of the jesters. These three styles relate definitely to the kind of dance and not to the status of the dancer. The *taupo's* dance is grave, aloof, beautiful. She is required to preserve a set, dreamy, nonchalant expression of infinite hauteur and detachment. The only permissible alternative to this expression is a series of grimaces, impudent rather than comic in nature and deriving their principal appeal from the strong contrast which they present to the more customary gravity. The *manaia* also when he dances in his *manaia* rôle is required to follow this same decorous and dignified pattern. Most little girls and a few little boys pattern their dancing on this convention. Chiefs, on the rare occasions when they consent to dance, and older women of rank have the privilege of choosing between this style and the adoption of a comedians rôle. The boys' dance is much jollier than the girls'. There is much greater freedom of movement and a great deal of emphasis on the noise made by giving rapid rhythmical slaps to the unclothed portions of the body which produce a crackling tattoo of sound. This style is neither salacious nor langorous although the *taupo's* dance is often both. It is athletic, slightly rowdy, exuberant, and owes much of its appeal to the feats of rapid and difficult co-ordination which the slapping involves. The jester's dance is peculiarly the dance of those who dance upon either side of the *taupo*, or the *manaia*, and honour them by mocking them. It is primarily the prerogative of talking chiefs and old men and women in general. The original motive is contrast; the jester provides comic relief for the stately dance of the *taupo*, and the higher the rank of the *taupo*, the higher the rank of the men and women who will condescend to act as clownish foils to her ability. The dancing of these jesters is characterized by burlesque, horseplay, exaggeration of the stereotyped figures, a great deal of noise made by hammering on the open mouth with spread palm, and a large amount of leaping about and pounding on the floor. The clown is occasionally so proficient that he takes the centre of the floor on these ceremonious occasions.

The little girl who is learning to dance has these three styles from which to choose, she has twenty-five or thirty figures from which to compose her dance and most important of all she has the individual dancers to watch. My first interpretation of the skill of the younger children was that they each took an older boy or girl as a model and sedulously and slavishly copied the whole dance. But I was not able to find a single instance in which a child would admit or seemed in any way conscious of having copied another; nor did I find, after closer familiarity with the group, any younger child whose style of dancing could definitely be referred to the imitation of another dancer. The style of every dancer of any virtuosity is known to every one in the village and when it is copied, it is copied conspicuously so that Vaitogi, the little girl who places her forearms parallel with the top of her head, her palms flat on her head, and advances in a stooping position, uttering hissing sounds, will be said to be dancing *a la Sina*. There is no stigma upon such imitation; the author does not resent it nor particularly glory in it; the crowd does not upbraid it; but so strong is the feeling for individualisation that a dancer will seldom introduce more than one such feature into an evening's performance; and when the dancing of two girls is similar, it is similar in spite of the efforts of both, rather than because of any attempt at imitation. Naturally, the dancing of the young children is much more similar than the dancing of the young men and girls who had had time and opportunity really to perfect a style.

The attitude of the elders towards precocity in singing, leading the singing or dancing, is in striking contrast to their attitude towards every other form of precocity. On the dance floor the dreaded accusation, "You are presuming above your age," is never heard. Little boys who would be rebuked and possibly whipped for such behaviour on any other occasion are allowed to preen themselves, to swagger and bluster and take the limelight without a word of reproach. The relatives crow with delight over a precocity for which they would hide their heads in shame were it displayed in any other sphere.

It is on these semi-formal occasions that the dance really

serves as an educational factor. The highly ceremonious dance of the *taupo* or *manaia* and their talking chiefs at a wedding or a *malaga*, with its elaborate costuming, compulsory distribution of gifts, and its vigilant attention to precedent and prerogative, offers no opportunities to the amateur or the child. They may only cluster outside the guest house and watch the proceedings. The existence of such a heavily stylized and elaborate archetype of course serves an additional function in giving zest as well as precedent to the informal occasions which partially ape its grandeur.

The significance of the dance in the education and socialisation of Samoan children is two-fold. In the first place it effectively offsets the rigorous subordination in which children are habitually kept. Here the admonitions of the elders change from "Sit down and keep still!" to "Stand up and dance!" The children are actually the centre of the group instead of its barely tolerated fringes. The parents and relatives distribute generous praise by way of emphasizing their children's superiority over the children of their neighbours or their visitors. The ubiquitous ascendency of age is somewhat relaxed in the interests of greater proficiency. Each child is a person with a definite contribution to make regardless of sex and age. This emphasis on individuality is carried to limits which seriously mar the dance as an æsthetic performance. The formal adult dance with its row of dancers, the *taupo* in the centre and an even number of dancers on each side focussed upon her with every movement directed towards accentuating her dancing, loses both symmetry and unity in the hands of the ambitious youngsters. Each dancer moves in a glorious individualistic oblivion of the others, there is no pretence of co-ordination or of subordinating the wings to the centre of the line. Often a dancer does not pay enough attention to her fellow dancers to avoid continually colliding with them. It is a genuine orgy of aggressive individualistic exhibitionism. This tendency, so blatantly displayed on these informal occasions, does not mar the perfection of the occasional formal dance when the solemnity of the occasion becomes a sufficient check upon the participants' aggressiveness. The formal dance is of personal significance only to people of rank or to the virtuoso to whom it presents a perfect occasion for display.

The second influence of the dance is its reduction of the threshold of shyness. There is as much difference between one Samoan child and another in the matter of shyness and self-consciousness as is apparent among our children, but where our shyest children avoid the limelight altogether, the Samoan child looks pained and anxious but dances just the same. The limelight is regarded as inevitable and the child makes at least a minimum of effort to meet its requirements by standing up and going through a certain number of motions. The beneficial effects of this early habituation to the public eye and the resulting control of the body are more noticeable in the case of boys than of girls. Fifteen- and sixteen-year-old boys dance with a charm and a complete lack of self-consciousness which is a joy to watch. The adolescent girl whose gawky, awkward gait and lack of co-ordination may be appalling, becomes a graceful, self-possessed person upon the dance floor. But this ease and poise does not seem to be carried over into everyday life with the same facility as it is in the case of young boys.

In one way this informal dance floor approximates more closely to our educational methods than does any other aspect of Samoan education. For here the precocious child is applauded, made much of, given more and more opportunities to show its proficiency while the stupid child is rebuked, neglected and pushed to the wall. This difference in permitted practice is reflected in increasing differences in the skill of the children as they grow older. Inferiority feeling in the classic picture which is so frequent in our society is rare in Samoa. Inferiority there seems to be derived from two sources, clumsiness in sex relations which affects the young men after they are grown and produces the *moetotolo*, and clumsiness upon the dance floor. I have already told the story of the little girl, shy beyond her fellows, whom prospective high rank had forced into the limelight and made miserably diffident and self-conscious.

And the most unhappy of the older girls was Masina, a girl about three years past puberty. Masina could not dance. Every one in the village knew that she could not dance. Her contemporaries deplored it; the younger children made fun of her. She had little charm, was deprecating in her manner, awkward, shy, and ill at ease. All of her five lovers had been casual, all temporary, all unimportant. She associated with girls much younger than herself. She had no self-confidence. No one sought her hand in marriage and she would not marry until her family needed the kind of property which forms a bride price.

It is interesting to notice that the one aspect of life in which the elders actively discriminate against the less proficient children seems to be the most powerful determinant in giving the children a feeling of inferiority.

The strong emphasis upon dancing does not discriminate against the physically defective. Instead every defect is capitalised in the form of the dance or compensated for by the perfection of the dance. I saw one badly hunchbacked boy who had worked out a most ingenious imitation of a turtle and also a combination dance with another boy in which the other supported him on his back. Ipu, the little albino, danced with aggressive facility and with much applause, while mad Laki, who suffered from a delusion that he was the high chief of the island, was only too delighted to dance for any one who addressed him with the elaborate courtesy phrases suitable to his rank. The dumb brother of the high chief of one village utilised his deaf mute gutturals as a running accompaniment to his dance, while the brothers of a fourteen-year-old feeble-minded mad boy were accustomed to deck his head with branches which excited him to a frenzied rhythmical activity, suggesting a stag whose antlers had been caught in the bush. The most precocious girl dancer in Taū was almost blind. So every defect, every handicap was included in this universal, specialized exploitation of personality.

The dancing child is almost always a very different person from her everyday self. After long acquaintance it is sometimes possible to guess the type of dance which a particular girl will do. This is particularly easy in the case of obviously tom-boy girls, but one is continually fooled by the depths of sophistication in the dancing of some pensive, dull child, or the lazy grace of some noisy little hoodlum.

Formal dancing displays are a recognised social entertainment and the highest courtesy a chief can offer his guest is to have his *taupo* dance for him. So likewise the boys dance after they have been tattooed, the *manaia* dances when he goes to woo his bride, the bride dances at her wedding. In the midnight conviviality of a *malaga* the dance often becomes flagrantly obscene and definitely provocative in character, but both of these are special developments of less importance than the function of informal dancing in the development of individuality and the compensation for repression of personality in other spheres of life.

9

The Attitude Towards Personality

THE EASE with which personality differences can be adjusted by a change of residence prevents the Samoans from pressing one another too hard. Their evaluations of personality are a curious mixture of caution and fatalism. There is one word *musu* which expresses unwillingness and intractability whether in the mistress who refuses to welcome a hitherto welcome lover, the chief who refuses to lend his kava bowl, the baby who won't go to bed, or the talking chief who won't go on a *malaga*. The appearance of a *musu* attitude is treated with almost superstitious respect. Lovers will prescribe formulæ for the treatment of a mistress, "lest she become *musu*," and the behaviour of the suppliant is carefully orientated in respect to this mysterious undesirability. The feeling seems to be not that one is dealing with an individual in terms of his peculiar preoccupations in order to assure a successful outcome of a personal relationship, appealing now to vanity, now to fear, now to a desire for power, but rather that one is using one or another of a series of potent practices to prevent a mysterious and widespread psychological phenomenon from arising. Once this attitude has appeared, a Samoan habitually gives up the struggle without more detailed inquiry and with a minimum of complaint. This fatalistic acceptance of an inexplicable attitude makes for an odd incuriousness about motives. The Samoans are not in the least insensitive to differences between people. But their full appreciation of these differences is blurred by their conception of an obstinate disposition, a tendency to take umbrage, irascibility, contrasuggestibility, and particular biases as just so many roads to one attitude—*musu*.

This lack of curiosity about motivation is furthered by the conventional acceptance of a completely ambiguous answer to any personal question. The most characteristic reply to any question about one's motivation is *Ta ilo*, "search me," sometimes made more specific by the addition of "I don't know."[*] This is considered to be an adequate and acceptable answer in ordinary conversation although its slight curtness bars it out from ceremonious occasions. So deep seated is the habit of using this disclaimer that I had to put a taboo upon its use by the children in order to get the simplest question answered directly. When this ambiguous rejoinder is combined with a statement that one is *musu*, the result is the final unrevealing statement, "Search me, why, I don't want to, that's all." Plans will be abandoned, children refuse to live at home, marriages broken off. Village gossip is interested in the fact but shrugs its shoulders before the motives.

There is one curious exception to this attitude. If an individual falls ill, the explanation is sought first in the attitudes of his relatives. Anger in the heart of a relative, especially in that of a sister, is most potent in producing evil and so the whole

[*] See Appendix I, page 148.

household is convened, a kava ceremony held and each relative solemnly enjoined to confess what anger there is in his heart against the sick person. Such injunctions are met either by solemn disclaimers or by detailed confessions: "Last week my brother came into the house and ate all the food, and I was angry all day"; or "My brother and I had a quarrel and my father took my brother's side and I was angry at my father for his favouritism towards my brother." But this special ceremony only serves to throw into strong relief the prevalent unspeculative attitude towards motivation. I once saw a girl leave a week-end fishing party immediately upon arrival at our destination and insist upon returning in the heat of the day the six miles to the village. But her companions ventured no hypothesis; she was simply *musu* to the party.

How great a protection for the individual such an attitude is will readily be seen when it is remembered how little privacy any one has. Chief or child, he dwells habitually in a house with at least half a dozen other people. His possessions are simply rolled in a mat, placed on the rafters or piled carelessly into a basket or a chest. A chief's personal property is likely to be respected, at least by the women of the household, but no one else can be sure from hour to hour of his nominal possessions. The tapa which a woman spent three weeks in making will be given away to a visitor during her temporary absence. The rings may be begged off her fingers at any moment. Privacy of possessions is virtually impossible. In the same way, all of an individual's acts are public property. An occasional love affair may slip through the fingers of gossip, and an occasional *moetotolo* go uncaught, but there is a very general cognisance on the part of the whole village of the activity of every single inhabitant. I shall never forget the outraged expression with which an informant told me that nobody, actually nobody at all, knew who was the father of Fa'amoana's baby. The oppressive atmosphere of the small town is all about them; in an hour children will have made a dancing song of their most secret acts. This glaring publicity is compensated for by a violent gloomy secretiveness. Where a Westerner would say, "Yes, I love him but you'll never know how far it went," a Samoan would say, "Yes, of course I lived with him, but you'll never know whether I love him or hate him."

The Samoan language has no regular comparative. There are several clumsy ways of expressing comparison by using contrast, "This is good and that is bad"; or by the location, "And next to him there comes, etc." Comparisons are not habitual although in the rigid social structure of the community, relative rank is very keenly recognised. But relative goodness, relative beauty, relative wisdom are unfamiliar formalisations to them. I tried over and over again to get judgments as to who was the wisest or the best man of the community. An informant's first impulse was always to answer: "Oh, they are all good"; or "There are so many wise ones." Curiously enough, there seemed to be less difficulty in distinguishing the vicious than the virtuous. This is probably due to the Missionary influence which if it has failed to give the native a conviction of Sin, has at least provided him with a list of sins. Although I often met with a preliminary response, "There are so many bad boys"; it was usually qualified spontaneously by "But so-and-so is the worst because he . . ." Ugliness and viciousness were more vivid and unusual attributes of personality; beauty, wisdom, and kindness were taken for granted.

In an account given of another person the sequence of traits mentioned followed a set and objective pattern: sex, age, rank, relationship, defects, activities. Spontaneous comment upon character or personality were unusual. So a girl describes her grandmother: "Lauuli? Oh, she is an old woman, very old, she's my father's mother. She's a widow with one eye. She is too old to go inland but sits in the house all day. She makes tapa." * This completely unanalytical account is only modified in the case of exceptionally intelligent adults who are asked to make judgments.

In the native classification attitudes are qualified by four terms, good and bad, easy and difficult, paired. A good child will be said to listen easily or to act well, a bad child to listen with difficulty or act badly. "Easy" and "with difficulty" are judgments of character; "good" and "bad" of behaviour. So that good or bad behaviour have become, explained in terms of ease or difficulty, to be regarded as an inherent capability of the individual. As we would say a person sang easily or swam without effort, the Samoan will say one obeys easily, acts respectfully, "easily," reserving the terms "good" or "well" for objective approbation. So a chief who was commenting on the bad behaviour of his brother's daughter remarked, "But Tui's children always did listen with difficulty," with as casual an acceptance of an irradicable defect as if he had said, "But John always did have poor eye sight."

Such an attitude towards conduct is paralleled by an equally unusual attitude towards the expression of emotion. The expressions of emotions are classified as "caused" and "uncaused." The emotional, easily upset, moody person is described as laughing without cause, crying without cause, show-

---

* For additional character sketches see Appendix I, page 149.

ing anger or pugnaciousness without cause. The expression "to be very angry without cause" does not carry the implication of quick temper, which is expressed by the word "to anger easily," nor the connotation of a disproportionate response to a legitimate stimulus, but means literally to be angry without cause, or freely, an emotional state without any apparent stimulus whatsoever. Such judgments are the nearest that the Samoan approaches to evaluation of temperament as opposed to character. The well-integrated individual who approximates closely to the attitudes of his age and sex group is not accused of laughing, crying, or showing anger without cause. Without inquiry it is assumed that he has good typical reasons for a behaviour which would be scrutinised and scorned in the case of the temperamental deviant. And always excessive emotion, violent preferences, strong allegiances are disallowed. The Samoan preference is for a middle course, a moderate amount of feeling, a discreet expression of a reasonable and balanced attitude. Those who care greatly are always said to care without cause.

The one most disliked trait in a contemporary is expressed by the term *fiasili,* literally "desiring to be highest," more idiomatically, "stuck up." This is the comment of the age mate where an older person would use the disapproving *tautala laititi,* "presuming above one's age." It is essentially the resentful comment of those who are ignored, neglected, left behind upon those who excel them, scorn them, pass them by. As a term of reproach it is neither as dreaded nor as resented as the *tautala laititi* because envy is felt to play a part in the taunt.

In the casual conversations, the place of idle speculation about motivation is taken by explanations in terms of physical defect or objective misfortune, thus "Sila is crying over in that house. Well, Sila is deaf." "Tulipa is angry at her brother. Tulipa's mother went to Tutuila last week." Although these statements have the earmarks of attempted explanations they are really only conversational habits. The physical defect or recent incident, is not specifically invoked but merely mentioned with slightly greater and more deprecatory emphasis. The whole preoccupation is with the individual as an actor, and the motivations peculiar to his psychology are left an unplumbed mystery.

Judgments are always made in terms of age groups, from the standpoints of the group of the speaker and the age of the person judged. A young boy will not be regarded as an intelligent or stupid, attractive or unattractive, clumsy or skilful person. He is a bright little boy of nine who runs errands efficiently and is wise enough to hold his tongue when his elders are present, or a promising youth of eighteen who can make excellent speeches in the *Aumaga,* lead a fishing expedition with discretion and treat the chiefs with the respect which is due to them, or a wise *matai,* whose words are few and well chosen and who is good at weaving eel traps. The virtues of the child are not the virtues of the adult. And the judgment of the speaker is similarly influenced by age, so that the relative estimation of character varies also. Pre-adolescent boys and girls will vote that boy and girl worst who are most pugnacious, irascible, contentious, rowdy. Young people from sixteen to twenty shift their censure from the rowdy and bully to the licentious, the *moetotolo* among the boys, the notoriously promiscuous among the girls; while adults pay very little attention to sex offenders and stress instead the inept, the impudent and the disobedient among the young, and the lazy, the stupid, the quarrelsome and the unreliable as the least desirable characters among the adults. When an adult is speaking the standards of conduct are graded in this fashion: small children should keep quiet, wake up early, obey, work hard and cheerfully, play with children of their own sex; young people should work industriously and skilfully, not be presuming, marry discreetly, be loyal to their relatives, not carry tales, nor be trouble makers; while adults should be wise, peaceable, serene, generous, anxious for the good prestige of their village and conduct their lives with all good form and decorum. No prominence is given to the subtler facts of intelligence and temperament. Preference between the sexes is given not to the arrogant, the flippant, the courageous, but to the quiet, the demure boy or girl who "speaks softly and treads lightly."

10

The Experience and Individuality of the Average Girl *

WITH A BACKGROUND of knowledge about Samoan custom, of the way in which a child is educated, of the claims which the community makes upon children and young people, of the attitude towards sex and personality, we come to the tale of the group of girls with whom I spent many months, the group of girls between ten and twenty years of age who lived in the three little villages on the lee side of the island of Taū. In their lives as a group, in their responses as individuals, lies the answer to the question: What is coming of age like in Samoa?

* See Tables and Summaries in Appendix IV.

The reader will remember that the principal activity of the little girls was baby-tending. They could also do reef fishing, weave a ball and make a pin-wheel, climb a cocoanut tree, keep themselves afloat in a swimming hole which changed its level fifteen feet with every wave, grate off the skin of a breadfruit or taro, sweep the sanded yard of the house, carry water from the sea, do simple washing and dance a somewhat individualised *siva*. Their knowledge of the biology of life and death was overdeveloped in proportion to their knowledge of the organisation of their society or any of the niceties of conduct prescribed for their elders. They were in a position which would be paralleled in our culture if a child had seen birth and death before she was taught not to pass a knife blade first or how to make change for a quarter. None of these children could speak the courtesy language, even in its most elementary forms, their knowledge being confined to four or five words of invitation and acceptance. This ignorance effectually barred them from the conversations of their elders upon all ceremonial occasions. Spying upon a gathering of chiefs would have been an unrewarding experience. They knew nothing of the social organisation of the village beyond knowing which adults were heads of families and which adult men and women were married. They used the relationship terms loosely and without any real understanding, often substituting the term, "sibling of my own sex," where a sibling of opposite sex was meant, and when they applied the term "brother" to a young uncle, they did so without the clarity of their elders who, while using the term in an age-grouping sense, realised perfectly that the "brother" was really a mother's or father's brother. In their use of language their immaturity was chiefly evidenced by a lack of familiarity with the courtesy language, and by much confusion in the use of the dual and of the inclusive and exclusive pronouns. These present about the same difficulty in their language as the use of a nominative after the verb "to be" in English. They had also not acquired a mastery of the processes for manipulating the vocabulary by the use of very freely combining prefixes and suffixes. A child will use the term *fa'a Samoa*, "in Samoan fashion," or *fa'atama*, tomboy, but fail to use the convenient *fa'a* in making a new and less stereotyped comparison, using instead some less convenient linguistic circumlocution.*

All of these children had seen birth and death. They had seen many dead bodies. They had watched miscarriage and peeked under the arms of the old women who were washing and commenting upon the undeveloped fœtus. There was no

* See Appendix I, page 153.

convention of sending children of the family away at such times, although the hordes of neighbouring children were scattered with a shower of stones if any of the older women could take time from the more absorbing events to hurl them. But the feeling here was that children were noisy and troublesome; there was no desire to protect them from shock or to keep them in ignorance. About half of the children had seen a partly developed fœtus, which the Samoans fear will otherwise be born as an avenging ghost, cut from a woman's dead body in the open grave. If shock is the result of early experiences with birth, death, or sex activities, it should surely be manifest here in this postmortem Cæsarian where grief for the dead, fear of death, a sense of horror and a dread of contamination from contact with the dead, the open, unconcealed operation and the sight of the distorted, repulsive fœtus all combine to render the experience indelible. An only slightly less emotionally charged experience was the often witnessed operation of cutting open any dead body to search out the cause of death. These operations performed in the shallow open grave, beneath a glaring noon-day sun, with a frighted, excited crowd watching in horrified fascination, are hardly orderly or unemotional initiations into the details of biology and death, and yet they seem to leave no bad effects on the children's emotional makeup. Possibly the adult attitude that these are horrible but perfectly natural, non-unique occurrences, forming a legitimate part of the child's experience, may sufficiently account for the lack of bad results. Children take an intense interest in life and death, and are more proportionately obsessed by it than are their adults who divide their horror between the death of a young neighbour in child-bed and the fact that the high chief has been insulted by some breach of etiquette in the neighbouring village. The intricacies of the social life are a closed book to the child and a correspondingly fascinating field of exploration in later life, while the facts of life and death are shorn of all mystery at an early age.

In matters of sex the ten-year-olds are equally sophisticated, although they witness sex activities only surreptitiously, since all expressions of affection are rigorously barred in public. A couple whose wedding night may have been spent in a room with ten other people will never the less shrink in shame from even touching hands in public. Individuals between whom there have been sex relations are said to be "shy of each other," and manifest this shyness in different fashion but with almost the same intensity as in the brother and sister avoidance. Husbands and wives never walk side by side through the village, for the husband, particularly, would be "ashamed." So no Samoan child is accustomed to seeing father and mother exchange casual caresses. The customary salutation by rubbing noses is, of course, as highly conventionalised and impersonal as our handshake. The only sort of demonstration which ever occurs in public is of the horseplay variety between young people whose affections are not really involved. This romping is particularly prevalent in groups of women, often taking the form of playfully snatching at the sex organs.

But the lack of privacy within the houses where mosquito netting marks off purely formal walls about the married couples, and the custom of young lovers of using the palm groves for their rendezvous, makes it inevitable that children should see intercourse, often and between many different people. In many cases they have not seen first intercourse, which is usually accompanied by greater shyness and precaution. With the passing of the public ceremony, defloration forms one of the few mysteries in a young Samoan's knowledge of life. But scouring the village palm groves in search of lovers is one of the recognised forms of amusement for the ten-year-olds.

Samoan children have complete knowledge of the human body and its functions, owing to the custom of little children going unclothed, the scant clothing of adults, the habit of bathing in the sea, the use of the beach as a latrine and the lack of privacy in sexual life. They also have a vivid understanding of the nature of sex. Masturbation is an all but universal habit, beginning at the age of six or seven. There were only three little girls in my group who did not masturbate. Theoretically it is discontinued with the beginning of heterosexual activity and only resumed again in periods of enforced continence. Among grown boys and girls casual homosexual practices also supplant it to a certain extent. Boys masturbate in groups but among little girls it is a more individualistic, secretive practice. This habit seems never to be a matter of individual discovery, one child always learning from another. The adult ban only covers the unseemliness of open indulgence.

The adult attitude towards all the details of sex is characterised by this view that they are unseemly, not that they are wrong. Thus a youth would think nothing of shouting the length of the village, "Ho, maiden, wait for me in your bed to-night," but public comment upon the details of sex or of evacuation were considered to be in bad taste. All the words which are thus banished from polite conversation are cherished by the children who roll the salacious morsels under their tongues with great relish. The children of seven and eight get as much illicit satisfaction out of the other functions of the body as out of sex. This is interesting in view of the different attitude in Samoa towards the normal processes of evacuation.

There is no privacy and no sense of shame. Nevertheless the brand of bad taste seems to be as effective in interesting the young children as is the brand of indecency among us. It is also curious that in theory and in fact boys and men take a more active interest in the salacious than do the women and girls.

It seems difficult to account for a salacious attitude among a people where so little is mysterious, so little forbidden. The precepts of the missionaries may have modified the native attitude more than the native practice. And the adult attitude towards children as nonparticipants may also be an important casual factor. For this seems to be the more correct view of any prohibitions which govern children. There is little evidence of a desire to preserve a child's innocence or to protect it from witnessing behaviour, the following of which would constitute the heinous offence, *tautala laititi* ("presuming above one's age"). For while a pair of lovers would never indulge in any demonstration before any one, child or adult, who was merely a spectator, three or four pairs of lovers who are relatives or friends often choose a common rendezvous. (This, of course, excludes relatives of opposite sex, included in the brother and sister avoidance, although married brothers and sisters might live in the same house after marriage.) From the night dances, now discontinued under missionary influence, which usually ended in a riot of open promiscuity, children and old people were excluded, as non-participants whose presence as uninvolved spectators would have been indecent. This attitude towards non-participants characterised all emotionally charged events, a women's weaving bee which was of a formal, ceremonial nature, a house-building, a candle-nut burning—these were activities at which the presence of a spectator would have been unseemly.

Yet, coupled with the sophistication of the children went no pre-adolescent heterosexual experimentation and very little homosexual activity which was regarded in native theory as imitative of and substitutive for heterosexual. The lack of precocious sex experimentation is probably due less to the parental ban on such precocity than to the strong institutionalised antagonism between younger boys and younger girls and the taboo against any amiable intercourse between them. This rigid sex dichotomy may also be operative in determining the lack of specialisation of sex feeling in adults. Since there is a heavily charged avoidance feeling towards brother and cousins, and a tendency to lump all other males together as the enemy who will some day be one's lovers, there are no males in a girl's age group whom she ever regards simply as individuals without relation to sex.

Such then was the experience of the twenty-eight little girls in the three villages. In temperament and character they varied enormously. There was Tita, who at nine acted like a child of seven, was still principally preoccupied with food, completely irresponsible as to messages and commissions, satisfied to point a proud fat finger at her father who was town crier. Only a year her senior was Pele, the precocious little sister of the loosest woman in the village. Pele spent most of her time caring for her sister's baby which, she delighted in telling you, was of disputed parentage. Her dancing in imitation of her sister's was daring and obscene. Yet, despite the burden of the heavy ailing baby which she carried always on her hip and the sordidness of her home where her fifty-year-old mother still took occasional lovers and her weak-kneed insignificant father lived a hen-pecked ignominious existence, Pele's attitude towards life was essentially gay and sane. Better than suggestive dancing she liked hunting for rare *samoana* shells along the beach or diving feet first into the swimming hole or hunting for land crabs in the moonlight. Fortunately for her, she lived in the centre of the Lumā gang. In a more isolated spot her unwholesome home and natural precocity might have developed very differently. As it was, she differed far less from the other children in her group than her family, the most notorious in the village, differed from the families of her companions. In a Samoan village the influence of the home environment is being continually offset in the next generation by group activities through which the normal group standards assert themselves. This was universally true for the boys for whom the many years' apprenticeship in the *Aumaga* formed an excellent school for disciplining individual peculiarities. In the case of the girls this function was formerly performed in part by the *Aualuma,* but, as I pointed out in the chapter on the girl and her age group, the little girl is much more dependent upon her neighbourhood than is the boy. As an adult she is also more dependent upon her relationship group.

Tuna, who lived next door to Pele, was in a different plight, the unwilling little victim of the great Samoan sin of *tautala laititi*. Her sister Lila had eloped at fifteen with a seventeen-year-old boy. A pair of hotheaded children, they had never thoroughly re-established themselves with the community, although their families had relented and solemnised the marriage with an appropriate exchange of property. Lila still smarted under the public disapproval of her precocity and lavished a disproportionate amount of affection upon her obstreperous baby whose incessant crying was the bane of the neighbourhood. After spoiling him beyond endurance, she would hand him over to Tuna. Tuna, a stocky little creature with a large head and enormous melting eyes, looked at life from a slightly oblique angle. She was a little more calculating than the other children, a little more watchful for returns, less given to gratuitous outlays of personal service. Her sister's overindulgence of the baby made Tuna's task much harder than those of her companions. But she reaped her reward in the slightly extra gentleness with which they treated their most burdened associate, and here again the group saved her from a pronounced temperamental response to the exigencies of her home life.

A little further away lived Fitu and Ula, Maliu and Pola, two pairs of sisters. Fitu and Maliu, girls of about thirteen, were just withdrawing from the gang, turning their younger brothers and sisters over to Ula and Pola, and beginning to take a more active part in the affairs of their households. Ula was alert, pretty, pampered. Her household might in all fairness be compared to ours; it consisted of her mother, her father, two sisters and two brothers. True, her uncle who lived next door was the *matai* of the household, but still this little biological family had a strong separate existence of its own and the children showed the results of it. Lalala, the mother, was an intelligent and still beautiful woman, even after bearing six children in close succession. She came from a family of high rank, and because she had had no brothers, her father had taught her much of the genealogical material usually taught to the favourite son. Her knowledge of the social structure of the community and of the minutiae of the ceremonies which had formerly surrounded the court of the king of Manu'a was as full as that of any middle-aged man in the community. She was skilled in the handicrafts and her brain was full of new designs and unusual applications of material. She knew several potent medical remedies and had many patients. Married at fifteen, while still a virgin, her marital life, which had begun with the cruel public defloration ceremony, had been her only sex experience. She adored her husband, whose poverty was due to his having come from another island and not to laziness or inability. Lalala made her choices in life with a full recognition of the facts of her existence. There was too much for her to do. She had no younger sisters to bear the brunt of baby-tending for her. There were no youths to help her husband in the plantations. Well and good, she would not wrestle with the inevitable. And so Lalala's house was badly kept. Her children were dirty and bedraggled. But her easy good nature did not fail her as she tried to weave a fine mat on some blazing afternoon, while the baby played with the brittle easily broken pandanus strands, and doubled her work. But all of this reacted upon Fitu, lanky, ill-favoured executive little creature that she was. Fitu combined a passionate devotion to her mother with an obsessive solicitude for her younger brothers and sisters. Towards Ula alone her attitude was mixed. Ula, fifteen months younger, was pretty, lithe, flexible and indolent. While Fitu was often teased by her mother and rebuked by her companions for being like a boy, Ula was excessively feminine. She worked as hard as any other child of her age, but Fitu felt that their mother and their home were unusual and demanded more than the average service and devotion. She and her mother were like a pair of comrades, and Fitu bossed and joked with her mother in a fashion shocking to all Samoan onlookers. If Fitu was away at night, her mother went herself to look for her, instead of sending another child. Fitu was the eldest daughter, with a precocity bred of responsibility and an efficiency which was the direct outcome of her mother's laissez-faire attitude. Ula showed equally clearly the effect of being the prettier younger sister, trading upon her superior attractiveness and more meagre sense of duty. These children, as did the children in all three of the biological families in the three villages, showed more character, more sharply defined personality, greater precocity and a more personal, more highly charged attitude towards their parents.

It would be easy to lay too much stress on the differences between children in large households and children in small ones. There were, of course, too few cases to draw any final conclusions. But the small family in Samoa *did* demand from the child the very qualities which were frowned upon in Samoan society, based upon the ideal of great households in which there were many youthful labourers who knew their place. And in these small families where responsibility and initiative were necessary, the children seemed to develop them much earlier than in the more usual home environment in which any display of such qualities was sternly frowned upon.

This was the case with Malui and Meta, Ipu and Vi, Mata, Tino and Lama, little girls just approaching puberty who lived in large heterogeneous households. They were giving over baby-tending for more productive work. They were reluctantly acquiring some of the rudiments of etiquette; they were slowly breaking their play affiliations with the younger children. But all of this was an enforced change of habits rather than any change in attitude. They were conscious of their new position as almost grown girls who could be trusted to go fishing or work on the plantations. Under their short dresses they again wore *lavalavas* which they had almost forgotten how to keep fastened. These dragged about their legs and cramped their movements and fell off if they broke into a sprint. Most of all they missed the gang life and eyed a little wistfully the activities of their younger relatives. Their large impersonal household provided them with no personal lives, invested them with no intriguing responsibilities. They were simply little girls who were robust enough to do heavy work and old enough to learn to do skilled work, and so had less time for play.

In general attitude, they differed not at all from Tolo, from Tulipa, from Lua, or Lata, whose first menstruation was a few months past. No ceremony had marked the difference between the two groups. No social attitude testified to a crisis past. They were told not to make kava while menstruating, but the participation in a restriction they'd known about all their lives was unimpressive. Some of them had made kava before puberty, others had not. It depended entirely upon whether there was an available girl or boy about when a chief wished to have some kava made. In more rigorous days a girl could not make kava nor marry until she menstruated. But the former restriction had yielded to the requirements of expediency. The menstruating girl experienced very little pain which might have served to stress for her her new maturity. All of the girls reported back or abdominal pains which, however, were so slight that they seldom interfered in any way with their usual activities. In the table I have counted it unusual pain whenever a girl was incapacitated for work, but these cases were in no sense comparable to severe cases of menstrual cramps in our civilisation. They were unaccompanied by dizziness, fainting spells, or pain sufficient to call forth groaning or writhing. The idea of such pain struck all Samoan women as bizarre and humorous when it was described to them. And no special solicitude for her health, mental or physical, was shown to the menstruating girl. From foreign medical advice they had learned that bathing during menstruation was bad, and a mother occasionally cautioned her daughter not to bathe. There was no sense of shame connected with puberty nor any need of concealment. Pre-adolescent children took the news that a girl had reached puberty, a woman had had a baby, a boat had come from Ofu, or a pig had been killed by a falling boulder with the same insouciance—all bits of diverting gossip; and any girl could give accurate testimony as to the development of any other girl in her neighbourhood or relationship groups. Nor was puberty the immediate forerunner of sex experience. Perhaps a year, two or even three years would pass before a girl's shyness would relax, or her figure appeal to the roving eye of some older boy. To be a virgin's first lover was considered the high point of pleasure and amorous virtuosity, so that a girl's first lover was usually not a boy of her own age, equally shy and inexperienced. The girls in this group were divided into little girls like Lua, and gawky overgrown Tolo, who said frankly that they did not want to go walking with boys, and girls like Pala, who while still virgins, were a little weary of their status and eager for amorous experience. That they remained in this passive untouched state so long was mainly due to the conventions of love-making, for while a youth liked to woo a virgin, he feared ridicule as a cradle-snatcher, while the girls also feared the dread accusation of *tautala laititi* ("presuming above one's age"). The forays of more seasoned middle-aged marauders among these very young girls were frowned upon, and so the adolescent girls were given a valuable interval in which to get accustomed to new work, greater isolation and an unfamiliar physical development.

The next older girls were definitely divided as to whether or not they lived in the pastor's households. A glance at the table in the appendix will show that among the girls a couple of years past puberty, there is a definite inverse correlation between residence at home and chastity, with only one exception, Ela, who had been forgiven and taken back into the household of a pastor where workers were short. Ela's best friend was her cousin, Talo, the only girl in the group who had sex experience before menstruation had begun. But Talo was clearly a case of delayed menstruation; all the other signs of puberty were present. Her aunt shrugged her shoulders in the face of Talo's obvious sophistication and winning charm and made no attempt to control her. The friendship between these two girls was one of the really important friendships in the whole group. Both girls definitely proclaimed their preference, and their homosexual practices were undoubtedly instrumental in producing Talo's precocity and solacing Ela for the stricter régime of the pastor's household.

These casual homosexual relations between girls never assumed any long-time importance. On the part of growing girls or women who were working together they were regarded as a pleasant and natural diversion, just tinged with the salacious. Where heterosexual relationships were so casual, so shallowly channelled, there was no pattern into which homosexual relationships could fall. Native theory and vocabulary recognised the real pervert who was incapable of normal heterosexual response, and the very small population is probably sufficient explanation for the rarity of these types. I saw only one, Sasi, a boy of twenty who was studying for the ministry. He was slightly but not pronouncedly feminine in appearance, was skilled at women's work and his homosexual drive was strong enough to goad him into making continual advances to other boys. He spent more time casually in the company of girls, maintained a more easy-going friendship with them than any other boy on the island. Sasi had proposed marriage to a girl in a pastor's household in a distant village and been refused, but as there was a rule that divinity students must marry before ordination, this has little significance. I could find no evidence that he had ever had heterosexual relations and the girls' casual attitude towards him was significant. They regarded him as an amusing freak while the men to whom he had made advances looked upon him with mingled annoyance and contempt. There were no girls who presented such a clear picture although three of the deviants discussed in the next chapter were clearly mixed types, without, however, showing convincing evidence of genuine perversion.

The general preoccupation with sex, the attitude that minor sex activities, suggestive dancing, stimulating salacious conversation, salacious songs and definitely motivated tussling are all acceptable and attractive diversions, is mainly responsible for the native attitude towards homosexual practices. They are simply *play*, neither frowned upon nor given much consideration. As heterosexual relations are given significance not by love and a tremendous fixation upon one individual, the only forces which can make a homosexual relationship lasting and important, but by children and the place of marriage in the economic and social structure of the village, it is easy to understand why very prevalent homosexual practices have no more important or striking results. The recognition and use in heterosexual relations of all the secondary variations of sex activity which loom as primary in homosexual relations are instrumental also in minimising their importance. The effects of chance childhood perversions, the fixation of attention on unusual erogenous zones with consequent transfer of sensitivity from the more normal centres, the absence of a definite and accomplished specialisation of erogenous zones—all the accidents of emotional development which in a civilisation, recognising only one narrow form of sex activity, result in unsatisfactory marriages, casual homosexuality and prostitution, are here rendered harmless. The Samoan puts the burden of amatory success upon the man and believes that women need more initiating, more time for the maturing of sex feeling. A man who fails to satisfy a woman is looked upon as a clumsy, inept blunderer, a fit object for village ridicule and contempt. The women in turn are conscious that their lovers use a definite technique which they regard with a sort of fatalism as if all men had a set of slightly magical, wholly irresistible, tricks up their sleeves. But amatory lore is passed down from one man to another and is looked upon much more self-consciously and analytically by men than by women. Parents are shy of going beyond the bounds of casual conversation (naturally these are much wider than in our civilisation) in the discussion of sex with their children, so that definite instruction passes from the man of twenty-five to the boy of eighteen rather than from father to son. The girls learn from the boys and do very little confiding in each other. All of a man's associates will know every detail of some unusual sex experience while the girl involved will hardly have confided the bare outlines to any one. Her lack of any confidants except relatives towards whom there is always a slight barrier of reserve (I have seen a girl shudder away from acting as an ambassador to her sister) may partly account for this.

The fact that educating one sex in detail and merely fortifying the other sex with enough knowledge and familiarity with sex to prevent shock produces normal sex adjustments is due to the free experimentation which is permitted and the rarity with which both lovers are amateurs. I knew of only one such case, where two children, a sixteen-year-old boy and a fifteen-year-old girl, both in boarding schools on another island, ran away together. Through inexperience they bungled badly. They were both expelled from school, and the boy is now a man of twenty-four with high intelligence and real charm, but a notorious *moetotolo,* execrated by every girl in his village. Familiarity with sex, and the recognition of a need of a technique to deal with sex as an art, have produced a scheme of personal relations in which there are no neurotic pictures, no frigidity, no impotence, except as the temporary result of severe illness, and the capacity for intercourse only once in a night is counted as senility.

Of the twenty-five girls past puberty, eleven had had heterosexual experience. Fala, Tolu, and Namu were three cousins who were popular with the youths of their own village and also with visitors from distant Fitiuta. The women of Fala's family were of easy virtue; Tolu's father was dead and she lived with her blind mother in the home of Namu's parents, who, burdened with six children under twelve years of age, were not going to risk losing two efficient workers by too close supervision. The three girls made common rendezvous with their lovers and their liaisons were frequent and gay. Tolu, the eldest, was a little weary after three years of casual adventures and professed herself willing to marry. She later moved into the household of an important chief in order to improve her chances of meeting strange youths who might be interested in matrimony. Namu was genuinely taken with a boy from Fitiuta whom she met in secret while a boy of her own village whom her parents favoured courted her openly. Occasional assignations with other boys of her own village relieved the monotony of life between visits from her preferred lover. Fala, the youngest, was content to let matters drift. Her lovers were friends and relatives of the lovers of her cousins and she was still sufficiently childlike and uninvolved to get almost as much enjoyment out of her cousins' love affairs as out of her own. All three of these girls worked hard, doing the full quota of work for an adult. All day they fished, washed, worked on the plantation, wove mats and blinds. Tolu was exceptionally clever at weaving. They were valuable economic assets to their families; they would be valuable to the husbands whom their families were not over anxious to find for them.

In the next village lived Luna, a lazy good-natured girl, three years past puberty. Her mother was dead. Her father had married again, but the second wife had gone back to her own people. Luna lived for several years in the pastor's household and had gone home when her stepmother left her father. Her father was a very old chief, tremendously preoccupied with his prestige and reputation in the village. He held an important title; he was a master craftsman; he was the best versed man in the village in ancient lore and details of ceremonial procedure. His daughter was a devoted and efficient attendant. It was enough. Luna tired of the younger girls who had been her companions in the pastor's household and sought instead two young married women among her relatives. One of these, a girl who had deserted her husband and was living with a temporary successor came to live in Luna's household. She and Luna were constant companions, and Luna, quite easily and inevitably took one lover, then two, then a third—all casual affairs. She dressed younger than her years, emphasised that she was still a girl. Some day she would marry and be a church member, but now: *Laititi a'u* ("I am but young"). And who was she to give up dancing.

Her cousin Lotu was a church member, and had attended the missionary boarding school. She had had only one accepted lover, the illegitimate son of a chief who dared not jeopardise his very slender chance of succeeding to his father's title by marrying her. She was the eldest of nine children, living in the third strictly biological family in the village. She showed the effects of greater responsibility at home by a quiet maturity and decision of manner, of her school training in a greater neatness of person and regard for the nicety of detail. Although she was transgressing, the older church members charitably closed their eyes, sympathising with her lover's family dilemma. Her only other sex experience had been with a *moetotolo*, a relative. Should her long fidelity to her lover lead to pregnancy, she would probably bear the child. (When a Samoan woman does wish to avoid giving birth to a child, exceedingly violent massage and the chewing of kava is resorted to, but this is only in very exceptional cases, as even illegitimate children are enthusiastically welcomed.) Lotu's attitudes were more considered, more sophisticated than those of the other girls of her age. Had it not been for the precarious social status of her lover, she would probably have been married already. As it was, she laboured over the care of her younger brothers and sisters, and followed the routine of relationship duties incumbent upon a young girl in the largest family on the island. She reconciled her church membership and her deviation from chastity by the tranquil reflection that she would have married had it been possible, and her sin rested lightly upon her.

In the household of one high chief lived the Samoan version of our devoted maiden aunts. She was docile, efficient, responsible, entirely overshadowed by several more attractive girls. To her were entrusted the new-born babies and the most difficult diplomatic errands. Hard work which she never resented took up all her time and energy. When she was asked to dance, she did so negligently. Others dancing so much more brilliantly, why make the effort? Hers was the appreciative worshipping disposition which glowed over Tolu's beauty or Fala's conquests or Alofi's new baby. She played the ukulele for others to dance, sewed flower necklaces for others to wear, planned rendezvous for others to enjoy, without humiliation or a special air of martyrdom. She admitted that she had had but one lover. He had come from far away; she didn't even know from what village, and he had never come back. Yes, probably she would marry some day if her chief so willed it, and was that the baby crying? She was the stuff of whom devoted aunts are made, depended upon and loved by all about her. A *malaga* to another village might have changed her life, for Samoa boys sought strange girls merely because they were strangers. But she was always needed at home by some one and younger girls went journeying in her stead.

Perhaps the most dramatic story was that of Moana, the last of the group of girls who lived outside the pastors' households, a vain, sophisticated child, spoiled by years of trading upon her older half-sister's devotion. Her amours had begun at fifteen and by the time a year and a half had passed, her parents, fearing that her conduct was becoming so indiscreet as to seriously mar her chances of making a good marriage, asked her uncle to adopt her and attempt to curb her waywardness. This uncle, who was a widower and a sophisticated rake, when he realised the extent of his niece's experience, availed himself also of her complacency. This incident, not common in Samoa, because of the great lack of privacy and isolation, would have passed undetected in this case, if Moana's older sister, Sila, had not been in love with the uncle also. This was the only example of prolonged and intense passion which I found in the three villages. Samoans rate romantic fidelity in terms of days or weeks at most, and are inclined to scoff at tales of life-long devotion. (They greeted the story of Romeo and Juliet with incredulous contempt.) But Sila was devoted to Mutu, her stepfather's younger brother, to the point of frenzy. She had been his mistress and still lived in his household, but his dilettantism had veered away from her indecorous intensity. When she discovered that he had lived with her sister, her fury knew no bounds. Masked under a deep solicitude for the younger girl, whom she claimed was an innocent untouched child, she denounced Mutu the length of the three villages. Moana's parents fetched her home again in a great rage and a family feud resulted. Village feeling ran high, but opinion was divided as to whether Mutu was guilty, Moana lying to cover some other peccadillo or Sila gossiping from spite. The incident was in direct violation of the brother and sister taboo for Mutu was young enough for Moana to speak of him as *tuagane* (brother). But when two months later, another older sister died during pregnancy, it was necessary to find some one stout-hearted enough to perform the necessary Cæsarian post-mortem operation. After a violent family debate, expediency triumphed and Mutu, most skilled of native surgeons, was summoned to operate on the dead body of the sister of the girl he had violated. When he later on announced his intention of marrying a girl from another island, Sila again displayed the most uncontrolled grief and despair, although she herself was carrying on a love affair at the time.

The lives of the girls who lived in the pastor's household differed from those of their less restricted sisters and cousins only in the fact that they had no love affairs and lived a more regular and ordered existence. For the excitement of moonlight trysts they substituted group activities, letting the pleasant friendliness of a group of girls fill their lesser leisure. Their interest in salacious material was slightly stronger than the interest of the girls who were free to experiment. They made real friends outside their relationship group, trusted other girls more, worked better in a group, were more at ease with one another but less conscious of their place in their own households than were the others.

With the exception of the few cases to be discussed in the next chapter, adolescence represented no period of crisis or stress, but was instead an orderly developing of a set of slowly maturing interests and activities. The girls' minds were perplexed by no conflicts, troubled by no philosophical queries, beset by no remote ambitions. To live as a girl with many lovers as long as possible and then to marry in one's own village, near one's own relatives and to have many children, these were uniform and satisfying ambitions.

11

The Girl in Conflict

WERE THERE NO CONFLICTS, no temperaments which deviated so markedly from the normal that clash was inevitable? Was the diffused affection and the diffused authority of the large families, the ease of moving from one family to another, the knowledge of sex and the freedom to experiment a sufficient guarantee to all Samoan girls of a perfect adjustment? In almost all cases, yes. But I have reserved for this chapter the tales of the few girls who deviated in temperament or in conduct, although in many cases these deviations were only charged with possibilities of conflict, and actually had no painful results.

The girl between fourteen and twenty stands at the centre of household pressure and can expend her irritation at her elders on those over whom she is in a position of authority. The possibility of escape seems to temper her restiveness under authority and the irritation of her elders also. When to the fear of a useful worker's running away is added also the fear of a daughter's indulging in a public elopement, and thus lowering her marriage value, any marked exercise of parental authority is considerably mitigated. Violent outbursts of wrath and summary chastisements do occur, but consistent and prolonged disciplinary measures are absent, and a display of temper is likely to be speedily followed by conciliatory measures. This, of course, applies only to the relation between a girl and her elders. Often conflicts of personality between young people of the same age in a household are not so tempered, but the removal of one party to the conflict, the individual with the weakest claims upon the household, is here also the most frequent solution. The fact that the age-group gang breaks up before adolescence and is never resumed except in a highly formal manner, coupled with the decided preference for household rather than group solidarity, accounts for the scarcity of conflict here. The child who shuns her age mates is more available for household work and is never worried by questions as to why she doesn't run and play with the other children. On the other hand, the tolerance of the children in accepting physical defect or slight strangeness of temperament prevents any child's suffering from unde-served ostracism.

The child who is unfavourably located in the village is the only real exile. Should the age group last over eight or ten years of age, the exiles would certainly suffer or very possibly as they grew bolder, venture farther from home. But the breakdown of the gang just as the children are bold enough and free enough to go ten houses from home, prevents either of these two results from occurring.

The absence of any important institutionalised relationship to the community is perhaps the strongest cause for lack of conflict here. The community makes no demands upon the young girls except for the occasional ceremonial service ren-dered at the meetings of older women. Were they delinquent in such duties it would be primarily the concern of their own households whose prestige would suffer thereby. A boy who refuses to attend the meetings of the *Aumaga*, or to join in the communal work, comes in for strong group disapproval and hostility, but a girl owes so small a debt to her community that it does not greatly concern itself to collect it.

The opportunity to experiment freely, the complete famil-iarity with sex and the absence of very violent preferences make her sex experiences less charged with possibilities of con-flict than they are in a more rigid and self-conscious civilisa-tion. Cases of passionate jealousy do occur but they are matters for extended comment and amazement. During nine months in the islands only four cases came to my attention, a girl who informed against a faithless lover accusing him of incest, a girl who bit off part of a rival's ear, a woman whose hus-band had deserted her and who fought and severely injured her successor, and a girl who falsely accused a rival of stealing. But jealousy is less expected and less sympathised with than among us, and consequently there is less of a pattern to which an individual may respond. Possibly conditions may also be simplified by the Samoan recognition and toleration of vindic-tive detraction and growling about a rival. There are no standards of good form which prescribe an insincere accept-ance of defeat, no insistence on reticence and sportsmanship. So a great deal of slight irritation can be immediately dis-sipated. Friendships are of so casual and shifting a nature that they give rise to neither jealousy nor conflict. Resentment is expressed by subdued grumblings and any strong resent-ment results in the angry one's leaving the household or some-times the village.

In the girl's religious life the attitude of the missionaries was the decisive one. The missionaries require chastity for church membership and discouraged church membership before marriage, except for the young people in the missionary boarding schools who could be continually supervised. This passive acceptance by the religious authorities themselves of pre-marital irregularities went a long way towards minimising the girls' sense of guilt. Continence became not a passport to heaven but a passport to the missionary schools which in turn were regarded as a social rather than a religious adventure. The girl who indulged in sex experiments was expelled from the local pastor's school, but it was notable that almost every other girl in the community, including the most notorious sex offenders, had been at one time resident of the pastors' households. The general result of the stricter supervision provided by these schools seemed to be to postpone the first sex experience two or three years. The seven girls in the household of one native pastor, the three in the household of the other, were all, although past puberty, living continent lives, in strong contrast to the habits of the rest of their age mates.

It might seem that there was fertile material for conflict between parents who wished their children to live in the pastor's house and children who did not wish to do so, and also between children who wished it and parents who did not.* This conflict was chiefly reduced by the fact that residence in the pastor's house actually made very little difference in the child's status in her own home. She simply carried her roll of mats, her pillow and her mosquito net from her home to the pastor's, and the food which she would have eaten at home was added to the quota of the food which her family furnished to the pastor. She ate her evening meal and slept at the pastor's; one or two days a week she devoted to working for the pastor's family, washing, weaving, weeding and sweeping the premises. The rest of her time she spent at home performing the usual tasks of a girl of her age, so that it was seldom that a parent objected strongly to sending a child to the pastor's. It involved no additional expense and was likely to reduce the chances of his daughter's conduct becoming embarrassing, to improve her mastery of the few foreign techniques, sewing, ironing, embroidery, which she could learn from the more skilled and schooled pastor's wife and thus increase her economic value.

If, on the other hand, the parents wished their children to stay and the children were unwilling to do so, the remedy was simple. They had but to transgress seriously the rules of the pastor's household, and they would be expelled; if they feared to return to their parents, there were always other relatives.

* See Appendix, page 151.

So the attitude of the church in respect to chastity held only the germs of a conflict which was seldom realised, because of the flexibility with which it adapted itself to the nearly inevitable. Attendance at the girls' main boarding school was an attractive prospect. The fascination of living in a large group of young people where life was easier and more congenial than at home, was usually a sufficient bribe to good behaviour, or at least to discretion. Confession of sin was a rare phenomenon in Samoa. The missionaries had made a rule that a boy who transgressed the chastity rule would be held back in his progress through the preparatory school and seminary for two years after the time his offence was committed. It had been necessary to change this ruling to read *two years from the detection of the offence,* because very often the offence was not detected until after the student had been over two years in the seminary, and under the old ruling, he would not have been punished at all. Had the young people been inspired with a sense of responsibility to a heavenly rather than an earthly decree and the boy or girl been answerable to a recording angel, rather than a spying neighbour, religion would have provided a real setting for conflict. If such an attitude had been coupled with emphasis upon church membership for the young and an expectation of religious experience in the lives of the young, crises in the lives of the young people would very likely have occurred. As it is, the whole religious setting is one of formalism, of compromise, of acceptance of half measure. The great number of native pastors with their peculiar interpretations of Christian teaching have made it impossible to establish the rigour of western Protestantism with its inseparable association of sex offences and an individual consciousness of sin. And the girls upon whom the religious setting makes no demands, make no demands upon it. They are content to follow the advice of their elders to defer church membership until they are older. *Laititi a'u. Fia siva* ("For I am young and like to dance"). The church member is forbidden to dance or to witness a large night dance. One of the three villages boasted no girl church members. The second village had only one, who had, however, long since transgressed her vows. But as her lover was a youth whose equivocal position in his family made it impossible to marry, the neighbours did not tattle where their sympathies were aroused, so Lotu remained tacitly a church member. In the third village there were two unmarried girls who were church members, Lita and Ana.

Lita had lived for years in the pastor's household and with one other girl, showed most clearly the results of a slightly alien environment. She was clever and executive, preferred the society of girls to that of boys, had made the best of her opportunities to learn English, worked hard at school, and wished to go to Tutuila and become a nurse or a teacher. Her ideals were thus just such as might frequently be found from any random selection of girls in a freshman class in a girls' college in this country. She coupled this set of individual ambitions with a very unusual enthusiasm for a pious father, and complied easily with his expressed wish for her to become a church member. After she left the pastor's household, she continued to go to school and apply herself vigorously to her studies, and her one other interest in life was a friendship with an older cousin who spoke some English and had had superior educational advantages in another island. Although this friendship had most of the trappings of a "crush" and was accompanied by the casual homosexual practices which are the usual manifestations of most associations between young people of the same sex, Lita's motivation was more definitely ambition, a desire to master every accessible detail of this alien culture in which she wished to find a place.

Sona, who was two years younger than Lita and had also lived for several years in the pastor's household, presented a very similar picture. She was overbearing in manner, arbitrary and tyrannous towards younger people, impudently deferential towards her elders. Without exceptional intellectual capacity she had exceptional persistence and had forced her way to the head of the school by steady dogged application. Lita, more intelligent and more sensitive, had left school for one year because the teacher beat her and Sona had passed above her, although she was definitely more stupid. Sona came from another island. Both her parents were dead and she lived in a large, heterogeneous household, at the beck and call of a whole series of relatives. Intent on her own ends, she was not enthusiastic about all this labour and was also unenthusiastic about most of her relatives. But one older cousin, the most beautiful girl in the village, had caught her imagination. This cousin, Manita, was twenty-seven and still unmarried. She had had many suitors and nearly as many lovers but she was of a haughty and aggressive nature and men whom she deemed worthy of her hand were wary of her sophisticated domineering manner. By unanimous vote she was the most beautiful girl in the village. Her lovely golden hair had contributed to half a dozen ceremonial headdresses. Her strategic position in her own family was heightened by the fact that her uncle, who had no hereditary right to make a *taupo*, had declared Manita to be his *taupo*. There was no other *taupo* in the village to dispute her claim. The murmurings were dying out; the younger children spoke of her as a *taupo* without suspicion; her beauty and ability as a dancer made it expedient to thus introduce her to visitors. Her family did not press her to marry, for the longer she remained unmarried, the stronger waxed the upstart legend. Her last lover had been a widower, a talking chief of intelligence and charm. He had loved Manita but he would not marry her. She lacked the docility which he demanded in a wife. Leaving Manita he searched in other villages for some very young girl whose manners were good but whose character was as yet unformed.

All this had a profound effect upon Sona, the ugly little stranger over whose lustreless eyes cataracts were already beginning to form. "Her sister" has no use for marriage; neither had she, Sona. Essentially unfeminine in outlook, dominated by ambition, she bolstered up her preference for the society of girls and a career by citing the example of her beautiful, wilful cousin. Without such a sanction she might have wavered in her ambitions, made so difficult by her already failing eyesight. As it was she went forward, blatantly proclaiming her pursuit of ends different from those approved by her fellows. Sona and Lita were not friends; the difference in their sanctions was too great; their proficiency at school and an intense rivalry divided them. Sona was not a church member. It would not have interfered with her behaviour in the least but it was part of her scheme of life to remain a school girl as long as possible and thus fend off responsibilities. So she, as often as the others, would answer, *Laititi a'u* ("I am but young"). While Lita attached herself to her cousin and attempted to learn from her every detail of another life, Sona identified herself passionately with the slightly more Europeanised family of the pastor, asserting always their greater relationship to the new civilisation, calling Ioane's wife, Mrs. Johns, building up a pitiable platform of *papalagi* (foreign) mannerisms as a springboard for future activities.

There was one other girl church member of Siufaga, Ana, a girl of nineteen. Her motives were entirely different. She was of a mild, quiescent nature, highly intelligent, very capable. She was the illegitimate child of a chief by a mother who had later married, run away, married again, been divorced, and finally gone off to another island. She formed no tie for Ana. Her father was a widower, living in a brother's house and Ana had been reared in the family of another brother. This family approximated to a biological one; there were two married daughters older than Ana, a son near her age, a daughter of fourteen and a crowd of little children. The father was a gentle, retiring man who had built his house outside the village, "to escape from the noise," he said. The two elder daughters married young and went away to live in their husbands' households. Ana and her boy cousin both lived in the pastor's household, while the next younger girl slept at home. The mother had a great distrust of men, especially of the young men of her own village. Ana should grow up to marry a pastor. She was not strong enough for the heavy work of the average Samoan wife. Her aunt's continuous harping on this strain, which was prompted mainly by a dislike of Ana's mother and a fear of the daughter's leaving home to follow in her mother's footsteps, had convinced Ana that she was a great deal too delicate for a normal existence. This theory received complete verification in the report of the doctor who examined the candidates for the nursing school and rejected her because of a heart murmur. Ana, influenced by her aunt's gloomy foreboding, was now convinced that she was too frail to bear children, or at least not more than one child at some very distant date. She became a church member, gave up dancing, clung closer to the group of younger girls in the pastor's school and to her foster home, the neurasthenic product of a physical defect, a small, isolated family group and the pastor's school.

These girls all represented the deviants from the pattern in one direction; they were those who demanded a different or improved environment, who rejected the traditional choices. At any time, they, like all deviants, might come into real conflict with the group. That they did not was an accident of environment. The younger girls in the pastor's group as yet showed fewer signs of being influenced by their slightly artificial environment. They were chaste where they would not otherwise have been chaste, they had friends outside their relationship group whom they would otherwise have viewed with suspicion, they paid more attention to their lessons. They still had not acquired a desire to substitute any other career for the traditional one of marriage. This was, of course, partly due to the fact that the pastor's school was simply one influence in their lives. The girls still spent the greater proportion of their waking time at home amid conventional surroundings. Unless a girl was given some additional stimulus, such as unusual home conditions, or possessed peculiarities of temperament, she was likely to pass through the school essentially unchanged in her fundamental view of life. She would acquire a greater respect for the church, a preference for slightly more fastidious living, greater confidence in other girls. At the same time the pastor's school offered a sufficient contrast to traditional Samoan life to furnish the background against which deviation could flourish. Girls who left the village and spent several years in the boarding school under the tutelage of white teachers were enormously influenced. Many of them became nurses; the majority married pastors, usually a deviation in attitude, involving as it did, acceptance of a different style of living.

So, while religion itself offered little field for conflict, the institutions promoted by religion might act as stimuli to new choices and when sufficiently reinforced by other conditions might produce a type of girl who deviated markedly from her companions. That the majority of Samoan girls are still unaffected by these influences and pursue uncritically the traditional mode of life is simply a testimony to the resistance of the native culture, which in its present slightly Europeanised state, is replete with easy solutions for all conflicts; and to the apparent fact that adolescent girls in Samoa do not generate their own conflicts, but require a vigorous stimulus to produce them.

These conflicts which have been discussed are conflicts of children who deviate upwards, who wish to exercise more choice than is traditionally permissible, and who, in making their choices, come to unconventional and bizarre solutions. The untraditional choices which are encouraged by the educational system inaugurated by the missionaries are education and the pursuit of a career and marriage outside of the local group (in the case of native pastors, teachers and nurses), preference for the society of one's own sex through prolonged and close association in school, a self-conscious evaluation of existence, and the consequent making of self-conscious choices. All of these make for increased specialisation, increased sophistication, greater emphasis upon individuality, where an individual makes a conscious choice between alternate or opposing lines of conduct. In the case of this group of girls, it is evident that the mere presentation of conflicting choices was not sufficient but that real conflict required the yeast of a need for choice and in addition a culturally favourable batter in which to work.

It will not be necessary to discuss another type of deviant, the deviant in a downward direction, or the delinquent. I am using the term delinquent to describe the individual who is maladjusted to the demands of her civilisation, and who comes definitely into conflict with her group, not because she adheres to a different standard, but because she violates the group standards which are also her own.*

* Such a distinction might well be made in the attitude towards delinquency in our own civilization. Delinquency cannot be defined even within one culture in terms of acts alone, but attitudes should also be considered. Thus the child who rifles her mother's purse to get money to buy food for a party or clothes to wear to a dance hall, who believes stealing is wrong, but cannot or will not resist the temptation to steal, is a delinquent, if the additional legal definition is given to her conduct by bringing her before some judicial authority. The young

A Samoan family or a Samoan community might easily come to conceive the conduct and standards of Sona and Lita as anti-social and undesirable. Each was following a plan of life which would not lead to marriage and children. Such a choice on the part of the females of any human community is, of course, likely to be frowned upon. The girls who, responding to the same stimuli, follow Sona's and Lita's example in the future will also run this risk.

But were there really delinquent girls in this little primitive village, girls who were incapable of developing new standards and incapable of adjusting themselves to the old ones? My group included two girls who might be so described, one girl who was just reaching puberty, the other a girl two years past puberty. Their delinquency was not a new phenomenon, but in both cases dated back several years. The members of their respective groups unhesitatingly pronounced them "bad girls," their age mates avoided them, and their relatives regretted them. As the Samoan village had no legal machinery for dealing with such cases, these are the nearest parallels which it is possible to drew with our "delinquent girl," substituting definite conflicts with unorganised group disapproval for the conflict with the law which defines delinquency in our society.

Lola was seventeen, a tall, splendidly developed, intelligent hoyden. She had an unusual endowment in her capacity for strong feeling, for enthusiasms, for violent responses to individuals. Her father had died when she was a child and she had been reared in a headless house. Her father's brother who was the *matai* had several houses and he had scattered his large group of dependents in several different parts of the village. So Lola, two older sisters, two younger sisters, and a brother a year older, were brought up by their mother, a kindly but ineffective woman. The eldest sister married and left the village when Lola was eight. The next sister, Sami, five years older than Lola, was like her mother, mild and gentle, with a soft undercurrent of resentment towards life running through all her quiet words. She resented and disliked her younger sister but she was no match for her. Nito, her brother, was a high-spirited and intelligent youth who might have

---

Christian communist who gives away her own clothes and also those of her brothers and sisters may be a menace to her family and to a society based upon private property, but she is not delinquent in the same sense. She has simply chosen an alternative standard. The girl who commits sex offenses with all attendant shame, guilt, and inability to defend herself from becoming continually more involved in a course of action which she is conscious is "wrong," until she becomes a social problem as an unmarried mother or a prostitute, is, of course, delinquent. The young advocate of free love who possesses a full quiver of ideals and sanctions for her conduct, may be undesirable, but from the standpoint of this discussion, she is not delinquent.

taught his sister a little wisdom had it not been for the brother and sister taboo which kept them always upon a formal footing. Aso, two years younger, was like Sami without Sami's sullen resentment. She adopted the plan of keeping out of Lola's way. The youngest, Siva, was like Lola, intelligent, passionate, easily aroused, but she was only eleven and merely profited by her sister's bad example. Lola was quarrelsome, insubordinate, impertinent. She contended every point, objected to every request, shirked her work, fought with her sisters, mocked her mother, went about the village with a chip upon her shoulder. When she was fourteen, she became so unmanageable at home that her uncle sent her to live in the pastor's houshold. She stayed there through a year of stormy scenes until she was finally expelled after a fight with Mala, the other delinquent. That she was not expelled sooner was out of deference to her rank as the niece of a leading chief. Her uncle realised the folly of sending her back to her mother. She was almost sixteen and well developed physically; and could be expected to add sex offences to the list of her troublesome activities at any moment. He took her to live in his own household under the supervision of his very strong-minded, executive wife, Pusa. Lola stayed there almost a year. It was a more interesting household than any in which she had lived. Her uncle's rank made constant calls upon her. She learned to make kava well, to dance with greater ease and mastery. A trip to Tutuila relieved the monotony of life; two cousins from another island came to visit, and there was much gaiety about the house. As consciousness of sex became more acute, she became slightly subdued and tentative in her manner. Pusa was a hard taskmaster and for a while Lola seemed to enjoy the novelty of a strong will backed by real authority. But the novelty wore off. The cousins prolonged their visit month after month. They persisted in treating her as a child. She became bored, sullen, jealous. Finally she ran away to other relatives, a very high chief's family, in the next village. Here, temporarily, was another house group of women folk, as the head of the house was in Tutuila, and his wife, his mother and his two children were the only occupants of the great guest house. Lola's labour was welcomed, and she set herself to currying favour with the high chief of the family. At first this was quite easy, as she had run away from the household of a rival chief and he appreciated her public defection. There were only much younger or much older girls in his household. Lola received the attention which she craved. The little girls resented her, but secretly admired her dashing uncompromising manner. But she had only been established here about a month when another chief, with a young and beautiful *taupo* in his train, came to visit her new chief and the whole party was lodged in the very house where she slept. Now began an endless round of hospitable tasks, and worst of all she must wait upon the pretty stranger who was a year younger than herself, but whose rank as visiting *taupo* gave her precedence. Lola again became troublesome. She quarrelled with the younger girls, was impertinent to the older ones, shirked her work, talked spitefully against the stranger. Perhaps all of this might have been only temporary and had no more far-reaching results than a temporary lack of favour in her new household, had it not been for a still more unfortunate event. The Don Juan of the village was a sleek, discreet man of about forty, a widower, a *matai*, a man of circumspect manner and winning ways. He was looking for a second wife and turned his attention toward the visitor who was lodged in the guest house of the next village. But Fuativa was a cautious and calculating lover. He wished to look over his future bride carefully and so he visited her house casually, without any declaration of his intention. And he noticed that Lola had reached a robust girlhood and stopped to pluck this ready fruit by the way, while he was still undecided about the more serious business of matrimony.

With all her capactiy for violence, Lola possessed also a strong capacity for affection. Fuativa was a skilled and considerate lover. Few girls were quite so fortunate in their first lovers, and so few felt such unmixed regret when the first love affair was broken off. Fuativa won her easily and after three weeks which were casual to him, and very important to her, he proposed for the hand of the visitor. The proposal itself might not have so completely enraged Lola although her pride was sorely wounded. Still, plans to marry a bride from such a great distance might miscarry. But the affianced girl so obviously demurred from the marriage that the talking chiefs became frightened. Fuativa was a rich man and the marriage ceremony would bring many perquisites for the talking chief. If the girl was allowed to go home and plead with her parents, or given the opportunity to elope with some one else, there would be no wedding perhaps and no rewards. The public defloration ceremony is forbidden by law. That the bridegroom was a government employé would further complicate his position should he break the law. So the anxious talking chief and the anxious suitor made their plans and he was given access to his future bride. The rage of Lola was unbounded and she took an immediate revenge, publicly accusing her rival of being a thief and setting the whole village by the ears. The women of the host household drove her out with many imprecations and she fled home to her mother, thus completing the residence cycle begun four years ago. She was now in the position of the delinquent in our society. She had continuously violated the group standards and she had exhausted all the solutions open to her. No other family group would open its doors to a girl whose record branded her as a liar, a trouble maker, a fighter, and a thief, for her misdeeds included continual petty thievery. Had she quarrelled with a father or been outraged by a brother-in-law, a refuge would have been easy to find. But her personality was essentially unfortunate. In her mother's household she made her sisters miserable, but she did not lord it over them as she had done before. She was sullen, bitter, vituperative. The young people of the village branded her as the possessor of a *lotu le aga*, ("a bad heart") and she had no companions. Her young rival left the island to prepare for her wedding, or the next chapter might have been Lola's doing her actual physical violence. When I left, she was living, idle, sullen, and defiant in her long-suffering mother's house.

Mala's sins were slightly otherwise. Where Lola was violent, Mala was treacherous; where Lola was antagonistic, Male was insinuating. Mala was younger, having just reached puberty in January, the middle of my stay on the island. She was a scrawny, ill-favoured little girl, always untidily dressed. Her parents were dead and she lived with her uncle, a sour, disgruntled man of small position. His wife came from another village and disliked her present home. The marriage was childless. The only other member of the house group was another niece who had divorced her husband. She also was childless. None showed Mala any affection, and they worked her unmercifully. The life of the only young girl or boy in a Samoan house, in the very rare cases when it occurs, is always very difficult. In this case it was doubly so. Ordinarily other relatives in the neighbourhood would have handed their babies over to her care, giving her a share in the activities of happier and more populous households. But from her early childhood she had been branded as a thief, a dangerous charge in a country where there are no doors or locks, and houses are left empty for a day at a time. Her first offence had been to steal a foreign toy which belonged to the chief's little son. The irate mother had soundly berated the child, on boat day, on the beach where all the people were gathered. When her name was mentioned, the information that she was a thief and a liar was tacked on as casually as was the remark that another was cross-eyed or deaf. Other children avoided her. Next door lived Tino, a dull good child, a few months younger than Mala. Ordinarily these two would have been companions and Mala always insisted that Tino was her friend, but Tino indignantly disclaimed all association with her. And as if her reputation for thievery were not sufficient, she added a further misdemeanor. She played with boys, preferred boys' games, tied her *lavalava* like a boy. This behaviour was displayed to the whole village who were vociferous in their condemnation. "She really was a very bad girl. She stole; she lied; and she played with boys." As in other parts of the world, the whole odium fell on the girl, so the boys did not fight shy of her. They teased her, bullied her, used her as general errand boy and fag. Some of the more precocious boys of her own age were already beginning to look to her for possibilities of other forms of amusement. Probably she will end by giving her favours to whoever asks for them, and sink lower and lower in the village esteem and especially in the opinion of her own sex from whom she so passionately desires recognition and affection.

Lola and Mala both seemed to be the victims of lack of affection. They both had unusual capacity for devotion and were abnormally liable to become jealous. Both responded with pathetic swiftness to any manifestations of affection. At one end of the scale in their need for affection, they were unfortunately placed at the other end in their chance of receiving it. Lola had a double handicap in her unfortunate temperament and the greater amiability of her three sisters. Her temperamental defects were further aggravated by the absence of any strong authority in her immediate household. Sami, the docile sister, had been saddled with the care of the younger children; Lola, harder to control, was given no such saving responsibility. These conditions were all as unusual as her demand and capacity for affection. And, similarly, seldom were children as desolate as Mala, marooned in a household of unsympathetic adults. So it would appear that their delinquency was produced by the combination of two sets of casual factors, unusual emotional needs and unusual home conditions. Less affectionate children in the same environments, or the same children in more favourable surroundings, probably would never have become as definitely outcast as these.

Only one other girl in the three villages calls for consideration under this conception of delinquency and she received far less general condemnation than either of the others. This was Sala, who lived in the third village. She lived in a household of seven, consisting of her widowed mother, her younger brother of ten, her grandmother, her uncle and his wife, and their two-year-old-son. This presented a fairly well-balanced family group and there were in addition many other relatives close by. Sala had been sent to live in the pastor's house but had speedily got involved in sex offenses and been expelled. Her attitude towards this pastor was still one of unveiled hostility. She was stupid, underhanded, deceitful and she possessed no aptitude for the simplest mechanical tasks. Her ineptness was the laughing stock of the village and her lovers were many and casual, the fathers of illegitimate children, men whose wives were temporarily absent, witless boys bent on a frolic. It was a saying among the girls of the village that Sala was apt at only one art, sex, and that she, who couldn't even sew thatch or weave blinds, would never get a husband. The social attitude towards her was one of contempt, rather than of antagonism, and she had experienced it keenly enough to have sunk very low in her own eyes. She had a sullen furtive manner, lied extravagantly in her assertions of skill and knowledge, and was ever on the alert for slights and possible innuendoes. She came into no serious conflict with her community. Her father beat her occasionally in a half-hearted manner, but her stupidity was her salvation for the Samoan possesses more charity towards weakness than towards misdirected strength. Sooner or later Sala's random sex experiences will probably lead to pregnancy, resulting in a temporary restriction of her activities and a much greater dependency upon her family. This economic dependence which in her case will be reinforced by her lack of manual skill will be strong enough to give her family a whip hand over her and force her to at least moderate her experimentation. She may not marry for many years and possibly will always be rated too inefficient for such responsibility.

The only delinquent in the making, that is a child who showed marked possibilities of increasing misbehaviour, was Siva, Lola's eleven-year-old little sister. She had the same obstreperous nature and was always engaging in fist fights with the other children, or hurling deadly insults after fleeing backs. She had the same violent craving for affection. But her uncle, profiting by her sister's unfortunate development, had taken her at the age of ten into his immediate family and so she was spending her pre-adolescent years under a much firmer régime than had her sister. And she differed from her sister in one respect, which was likely to prove her salvation. Where Lola had no sense of humour and no lightness of touch, Siva had both. She was a gifted mimic, an excruciatingly funny dancer, a born comedian. People forgave her her violence and her quarrelsomeness for sheer mirth over her propitiatory antics. If this facility continues to endear her to her aunts and cousins, who already put up with any number of pranks and fits of temper from her, she will probably not follow in her sister's steps. One affectionate word makes her shift her attention, and she has a real gift for affection. Once at a dancing party I had especially requested the children to be good and not waste time in endless bickerings and jealousies. I selected three little girls, the traditional number, to dance, and one of them, Meta, claimed that she had a sore foot. I turned hastily to Siva and asked her to fill out the figure. She was preparing to do so, with none too good grace at being second choice, when Meta, who had merely been holding back for more urging, leaped to her feet, and took the empty place. Siva was doubling up her fists ready to fly at Meta's throat when she caught my eye. She swallowed furiously, and then jerked the flower wreath from around her own neck and flung it over Meta's head. With better luck than her sister, she will not come into lasting conflict with her society.

And here ends the tale of serious conflict or serious deviation from group standards. The other girls varied as to whether they were subjected to the superior supervision of the pastor's household or not, as to whether they came from households of rank or families of small prestige, and most of all as to whether they lived in a biological family or a large heterogeneous household. But with differences in temperament equal to those found among us, though with a possibly narrower range of intellectual ability, they showed a surprising uniformity of knowledge, skill and attitude, and presented a picture of orderly, regular development in a flexible, but strictly delimited, environment.

12

Maturity and Old Age

BECAUSE THE COMMUNITY makes no distinction between unmarried girls and the wives of untitled men in the demands which it makes upon them, and because there is seldom any difference in sex experience between the two groups, the dividing line falls not between married and unmarried but between grown women and growing girls in industrial activity and between the wives of *matais* and their less important sisters in ceremonial affairs. The girl of twenty-two or twenty-three who is still unmarried loses her laissez-faire attitude. Family pressure is an effective cause in bringing about this change. She is an adult, as able as her married sisters and her brothers' young wives; she is expected to contribute as heavily as they to household undertakings. She lives among a group of contemporaries upon whom the responsibilities of marriage are making increased demands. Rivalry and emulation enter in. And also she may be becoming a little anxious about her own marital chances. The first preoccupation with sex experimentation has worn itself out and she settles down to increase her value as a wife. In native theory a girl knows how to sew thatch, but doesn't really make thatch until she is married. In actual practice the adult unmarried girls perform household and agricultural tasks identical with those performed by their married sisters, except that whereas pregnancy and nursing children tie the young married women to the house, the unmarried girls are free to go off on long fishing expeditions, or far inland in search of weaving materials.

A married couple may live either in the household of the girl or of the boy, choice being made on the basis of rank, or the industrial needs of the two households. The change of residence makes much less difference to the girl than to the boy. A married woman's life is lived in such a narrow sphere that her only associates are the women of her household. Residence in her husband's village instead of her own does not narrow her life, for her participation in village affairs will remain slight and unimportant until her husband assumes a title which confers status upon her also. If her husband's household is in her own village, her responsibilities will be increased somewhat because she will be subject to continual demands from her own near relatives as well as from those of her husband.

There is no expectation of conflict between daughter-in-law and mother-in-law. The mother-in-law must be respected because she is an elder of the household and an insolent daughter-in-law is no more tolerated than an insubordinate daughter or niece. But tales of the traditional lack of harmony which exists in our civilisation were treated by the Samoans with contemptuous amusement. Where the emotional ties between parents and children are so weak, it was impossible to make them see it as an issue between a man's mother and man's wife, in which jealousy played a part. They saw it simply as failure on the part of the young and unimportant person to pay proper respect to the old, granting of course that there were always irascible old people from whom it was expedient to move away. The same thing holds true for the young man, if he goes to live in his father-in-law's house. If the father-in-law is the *matai,* he has complete authority over his daughter's husband; if he is only an untitled old man, he must still be treated with respect.

But change of village for the young man makes a great difference, because he must take his place in a new *Aumaga,* and work with strangers instead of with the boys with whom he has worked and played since childhood. Very often he never becomes as thoroughly assimilated to the new group as he was to the old. He stands more upon his dignity. He works with his new companions but does not play with them. The social life of the *Aumaga* centres about the group courtesies which they pay to visiting girls. In his own village a man will accompany the younger boys on these occasions for many years after he is married. But in his wife's village, such behaviour becomes suddenly less appropriate. Random amatory adventures are also more hazardous when he is living as a member of his wife's household. And although his transition from the status of a young man to the status of a *matai* is easier, he ages more quickly; although he may earn great respect in his adopted village, he commands less of its affection.

In most marriages there is no sense of setting up a new and separate establishment. The change is felt in the change of residence for either husband or wife and in the reciprocal relations which spring up between the two families. But the young couple live in the main household, simply receiving a bamboo pillow, a mosquito net and a pile of mats for their bed. Only for the chief or the chief's son is a new house built. The wife works with all the women of the household and waits upon all the men. The husband shares the enterprises of the other men and boys. Neither in personal service given or received are the two marked off as a unit. Nor does marriage of either brother or sister slacken the avoidance rules; it merely adds another individual, the new sister or brother-in-law, to whom the whole series of avoidances must be applied. In the sexual relation alone are the two treated as one. For even in the care of the young children and in the decisions as to their future, the uncles and aunts and grandparents participate as fully as the parents. It is only when a man is *matai* as well as father, that he has control over his own children; and when this is so, the relationship is blurred in opposite fashion, for he has the same control over many other young people who are less closely related to him.

The pregnant young wife is surrounded by a multitude of taboos, most of which are prohibitions against solitary activities. She must not walk alone, sit alone, dance alone, gather food alone, eat alone, or when only her husband is present. All of these taboos are explained by the amiable doctrine that only things which are wrong are done in solitude and that any wrong deed committed by the expectant mother will injure the child. It seems simpler to prohibit solitary acts than wrong ones.

There are also ghosts which are particularly likely to injure the pregnant woman, and she is warned against walking in ghost-ridden places. She is warned against doing too heavy work and against getting chilled or overheated. While pregnancy is not treated with anything like the consideration which is often given it here, her first pregnancy gives a woman a certain amount of social prominence. This prominence is in direct proportion to her rank, and the young wife whose child is the presumptive heir to some high title is watched over with great solicitude. Relatives gather from great distances for the confinement and birth feast, which is described as the mother's feast, rather than the feast in honour of either child or father.

After the birth of the first child, the other children arrive frequently and with small remark. Old gossips count them and comment on the number living, dead or miscarried in previous births. A pig is roasted for the birth feast to which only the near relatives are invited. The mother of many children is rather taken for granted than praised. The barren woman is mildly execrated and her misfortune attributed to loose living. There were three barren older women on Taū; all three were midwives and reputed to be very wise. Now well past the child-bearing age, they were reaping the reward of the greater application to the intricacies of their calling with which they had compensated for their barrenness.

The young married women of twenty to thirty are a busy, cheerful group. They become church members and wear hats to church. When they have not a baby at the breast, they are doing heavy work on the plantations, fishing or making tapa. No other important event will ever happen to them again. If their husbands die, they will probably take new husbands, and those of lower rank. If their husbands become *matais*, they will also acquire a place in the *Fono* of the women. But it is only the woman with a flair for political wire-pulling and the luck to have either important relatives or an important husband who gets any real satisfaction out of the social organisation of the village.

The young men do not settle as early into a groove. What her first child is to a woman his title is to a man, and while each new child is less of an event in her life, a new title is always a higher one and a greater event in his. A man rarely attains his first title before he is thirty, often not before he is forty. All the years between his entrance into the *Aumaga* and his entrance into the *Fono* are years of striving. He cannot acquire a reputation and then rest upon it or another claimant to the same title will take advantage of his indolence and pass him in the race. One good catch of fish does not make him a fisherman nor one housebeam neatly adzed, a carpenter;

the whole emphasis is upon a steady demonstration of increasing skill which will be earnest of the necessary superiority over his fellows. Only the lazy, the shiftless, the ambitionless fail to respond to this competition. The one exception to this is in the case of the son or heir of the high chief who may be made the *manaia* at twenty. But here his high rank has already subjected him to more rigorous discipline and careful training than the other youths, and, as *manaia*, he is the titular head of the *Aumaga*, and must lead it well or lose his prestige.

Once having acquired a *matai* name and entered the *Fono*, differences in temperament prevail. The *matai* name he receives may be a very small one, carrying with it no right to a post in the council house, or other prerogatives. It may be so small that *matai* though he is, he does not try to command a household, but lives instead in the shadow of some more important relative. But he will be a member of the *Fono*, classed with the elders of the village, and removed forever from the hearty group activities of the young men. Should he become a widower and wish to court a new wife, he can only do so by laying aside his *matai* name and entering her house under the fiction that he is still a youth. His main preoccupation is the affairs of the village; his main diversion, hours spent in ceremonious argument in some meeting. He always carries his bundle of beaten cocoanut fibre and as he talks, he rolls the fibres together on his bare thigh.

The less ambitious rest upon this achievement. The more ambitious continue the game, for higher titles, for greater prestige as craftsmen or orators, for the control of more strings in the political game. At last the preference for the most able, the very preference which, in defiance of laws of primogeniture or direct descent, may have given a man his title, takes it away from him. For should he live beyond his prime, fifty-five or sixty, his name is taken from him and given to another, and he is given a "little *matai* name," so that he may still sit with the other *matais* and drink his kava. These old men stay at home, guard the house while the others go inland to the plantations, superintend the children, braid cinet and give advice, or in a final perverse assertion of authority, fail to give it. One young chief who had been given his father's name during his father's lifetime, complained to me: "I had no old man to help me. My father was angry that his title was given to me and he would tell me nothing. My mother was wise but she came from another island and did not know well the ancient ways of our village. There was no old one in the house to sit with me in the evening and fill my ears with the things from the olden time. A young *matai* should always have an old man beside him, who, even though he is deaf and cannot always hear his questions, can still tell him many things."

The women's lives pursue a more even tenor. The wives of chiefs and talking chiefs have to give some time to the mastery of ceremonial. The old women who become midwives or doctors pursue their professions but seldom and in a furtive, private fashion. The menopause is marked by some slight temperamental instability, irritability, finickiness about food, a tendency to sudden whims and inexplicable fancies. Once past the menopause and relieved of child-bearing, a woman turns her attention again to the heavy work of the plantations. The hardest work of the village is done by women between forty-five and fifty-five. Then, as age approaches, she settles down to performing the skilled tasks in the household, to weaving and tapa making.

Where a man is disqualified from active work by rheumatism, elephantiasis, or general feebleness, his rôle as a teacher is diminished. He can teach the aspirant young fisherman the lore of fishing but not the technique. The old woman on the other hand is mistress of housebound crafts and to her must go the girl who is ambitious to become a skilled weaver. Another can gather the herbs which she needs for her medicines, while she keeps the secret of compounding them. The ceremonial burning of the candle-nut to obtain black dye is in the hands of very old women. And also these old women are usually more of a power within the household than the old men. The men rule partly by the authority conferred by their titles, but their wives and sisters rule by force of personality and knowledge of human nature. A life-long preoccupation within the smaller group makes them omniscient and tyrannical. They suffer no diminution of prestige except such as is inherent in the complete loss of their faculties.

The feeling for generation is retained until death, and the very old people sit in the sun and talk softly without regard for taboo or sex.

13

Our Educational Problems in the Light of Samoan Contrasts

FOR MANY CHAPTERS we have followed the lives of Samoan girls, watched them change from babies to baby-tenders, learn to make the oven and weave fine mats, forsake the life of the gang to become more active members of the household, defer marriage through as many years of casual love-making as possible, finally marry and settle down to rearing children who will repeat the same cycle. As far as our material permitted, an experiment has been conducted to discover what the process of development was like in a society very different from our own. Because the length of human life and the complexity of our society did not permit us to make our experiment here, to choose a group of baby girls and bring them to maturity under conditions created for the experiment, it was necessary to go instead to another country where history had set the stage for us. There we found girl children passing through the same process of physical development through which our girls go, cutting their first teeth and losing them, cutting their second teeth, growing tall and ungainly, reaching puberty with their first menstruation, gradually reaching physical maturity, and becoming ready to produce the next generation. It was possible to say: Here are the proper conditions for an experiment; the developing girl is a constant factor in America and in Samoa; the civilisation of America and the civilisation of Samoa are different. In the course of development, the process of growth by which the girl baby becomes a grown woman, are the sudden and conspicuous bodily changes which take place at puberty accompanied by a development which is spasmodic, emotionally charged, and accompanied by an awakened religious sense, a flowering of idealism, a great desire for assertion of self against authority—or not? Is adolescence a period of mental and emotional distress for the growing girl as inevitably as teething is a period of misery for the small baby? Can we think of adolescence as a time in the life history of every girl child which carries with it symptoms of conflict and stress as surely as it implies a change in the girl's body?

Following the Samoan girls through every aspect of their lives we have tried to answer this question, and we found throughout that we had to answer it in the negative. The adolescent girl in Samoa differed from her sister who had not reached puberty in one chief respect, than in the older girl certain bodily changes were present which were absent in the younger girl. There were no other great differences to set off the group passing through adolescence from the group which would become adolescent in two years or the group which had become adolescent two years before.

And if one girl past puberty is undersized while her cousin is tall and able to do heavier work, there will be a difference between them, due to their different physical endowment, which will be far greater than that which is due to puberty. The tall, husky girl will be isolated from her companions, forced to do longer, more adult tasks, rendered shy by a change of clothing, while her cousin, slower to attain her growth, will still be treated as a child and will have to solve only the slightly fewer problems of childhood. The precedent of educators here who recommend special tactics in the treatment of adolescent girls translated into Samoan terms would read: Tall girls are different from short girls of the same age, we must adopt a different method of educating them.

But when we have answered the question we set out to answer we have not finished with the problem. A further question presents itself. If it is proved that adolescence is not necessarily a specially difficult period in a girl's life—and proved it is if we can find any society in which that is so—then what accounts for the presence of storm and stress in American adolescents? First, we may say quite simply, that there must be something in the two civilisations to account for the difference. If the same process takes a different form in two different environments, we cannot make any explanations in terms of the process, for that is the same in both cases. But the social environment is very different and it is to it that we must look for an explanation. What is there in Samoa which is absent in America, what is there in America which is absent in Samoa, which will account for this difference?

Such a question has enormous implications and any attempt to answer it will be subject to many possibilities of error. But if we narrow our question to the way in which aspects of Samoan life which irremediably affect the life of the adolescent girl differ from the forces which influence our growing girls, it is possible to try to answer it.

The background of these differences is a broad one, with two important components; one is due to characteristics which are Samoan, the other to characteristics which are primitive.

The Samoan background which makes growing up so easy, so simple a matter, is the general casualness of the whole society. For Samoa is a place where no one plays for very high stakes, no one pays very heavy prices, no one suffers for his convictions or fights to the death for special ends. Disagreements between parent and child are settled by the child's moving across the street, between a man and his village by the man's removal to the next village, between a husband and his wife's seducer by a few fine mats. Neither poverty nor great disasters threaten the people to make them hold their lives dearly and tremble for continued existence. No implacable gods, swift to anger and strong to punish, disturb the even tenor of their days. Wars and cannibalism are long since passed away and now the greatest cause for tears, short of death itself, is a journey of a relative to another island. No one is hurried along in life or punished harshly for slowness of development. Instead the gifted, the precocious, are held back, until the slowest among them have caught the pace. And in personal relations, caring is as slight. Love and hate, jealousy and revenge, sorrow and bereavement, are all matters of weeks. From the first months of its life, when the child is handed carelessly from one woman's hands to another's, the lesson is learned of not caring for one person greatly, not setting high hopes on any one relationship.

And just as we may feel that the Occident penalises those unfortunates who are born into Western civilisation with a taste for meditation and a complete distaste for activity, so we may say that Samoa is kind to those who have learned the lesson of not caring, and hard upon those few individuals who have failed to learn it. Lola and Mala and little Siva, Lola's sister, all were girls with a capacity for emotion greater than their fellows. And Lola and Mala, passionately desiring affection and too violently venting upon the community their disappointment over their lack of it, were both delinquent, unhappy misfits in a society which gave all the rewards to those who took defeat lightly and turned to some other goal with a smile.

In this casual attitude towards life, in this avoidance of conflict, of poignant situations, Samoa contrasts strongly not only with America but also with most primitive civilisations. And however much we may deplore such an attitude and feel that important personalities and great art are not born in so shallow a society, we must recognise that here is a strong factor in the painless development from childhood to womanhood. For where no one feels very strongly, the adolescent will not be tortured by poignant situations. There are no such disastrous choices as those which confronted young people who felt that the service of God demanded forswearing the world forever, as in the Middle Ages, or cutting off one's finger as a religious offering, as among the Plains Indians. So, high up in our list of explanations we must place the lack of deep feeling which the Samoans have conventionalised until it is the very framework of all their attitudes toward life.

And next there is the most striking way in which all isolated primitive civilisation and many modern ones differ from our own, in the number of choices which are permitted to each individual. Our children grow up to find a world of choices dazzling their unaccustomed eyes. In religion they may be Catholics, Protestants, Christian Scientists, Spiritualists, Agnostics, Atheists, or even pay no attention at all to religion. This is an unthinkable situation in any primitive society not exposed to foreign influence. There is one set of gods, one accepted religious practice, and if a man does not believe, his only recourse is to believe less than his fellows; he may scoff but there is no new faith to which he may turn. Present-day Manu'a approximates this condition; all are Christians of the same sect. There is no conflict in matters of belief although there is a difference in practice between Church-members and non-Church-members. And it was remarked that in the case of several of the growing girls the need for choice between these two practices may some day produce a conflict. But at present the Church makes too slight a bid for young unmarried members to force the adolescent to make any decision.

Similarly, our children are faced with half a dozen standards of morality: a double sex standard for men and women, a single standard for men and women, and groups which advocate that the single standard should be freedom while others argue that the single standard should be absolute monogamy. Trial marriage, companionate marriage, contract marriage—all these possible solutions of a social impasse are paraded before the growing children while the actual conditions in their own communities and the moving pictures and magazines inform them of mass violations of every code, violations which march under no banners of social reform.

The Samoan child faces no such dilemma. Sex is a natural, pleasurable thing; the freedom with which it may be indulged in is limited by just one consideration, social status. Chiefs' daughters and chiefs' wives should indulge in no extra-marital experiments. Responsible adults, heads of households and mothers of families should have too many important matters on hand to leave them much time for casual amorous adventures. Every one in the community agrees about the matter, the only dissenters are the missionaries who dissent so vainly that their protests are unimportant. But as soon as a sufficient sentiment gathers about the missionary attitude with its European standard of sex behaviour, the need for choice, the forerunner of conflict, will enter into Samoan society.

Our young people are faced by a series of different groups which believe different things and advocate different practices, and to each of which some trusted friend or relative may belong. So a girl's father may be a Presbyterian, an imperialist, a vegetarian, a teetotaler, with a strong literary preference for Edmund Burke, a believer in the open shop and a high tariff, who believes that woman's place is in the home, that young girls should wear corsets, not roll their stockings, not smoke, nor go riding with young men in the evening. But her mother's father may be a Low Episcopalian, a believer in high living, a strong advocate of States' Rights and the Monroe Doctrine, who reads Rabelais, likes to go to musical shows and horse races. Her aunt is an agnostic, an ardent advocate of woman's rights, an internationalist who rests all her hopes on Esperanto, is devoted to Bernard Shaw, and spends her spare time in campaigns of anti-vivisection. Her elder brother, whom she admires exceedingly, has just spent two years at Oxford. He is an Anglo-Catholic, an enthusiast concerning all things mediæval, writes mystical poetry, reads Chesterton, and means to devote his life to seeking for the lost secret of mediæval stained glass. Her mother's younger brother is an engineer, a strict materialist, who never recovered from reading Haeckel in his youth; he scorns art, believes that science will save the world, scoffs at everything that was said and thought before the nineteenth century, and ruins his health by experiments in the scientific elimination of sleep. Her mother is of a quietistic frame of mind, very much interested in Indian philosophy, a pacifist, a strict non-participator in life, who in spite of her daughter's devotion to her will not make any move to enlist her enthusiasms. And this may be within the girl's own household. Add to it the groups represented, defended, advocated by her friends, her teachers, and the books which she reads by accident, and the list of possible enthusiasms, of suggested allegiances, incompatible with one another, becomes appalling.

The Samoan girl's choices are far otherwise. Her father is a member of the Church and so is her uncle. Her father lives in a village where there is good fishing, her uncle in a village where there are plenty of cocoanut crabs. Her father is a good fisherman and in his house there is plenty to eat; her uncle is a talking chief and his frequent presents of bark cloth provide excellent dance dresses. Her paternal grandmother, who lives with her uncle, can teach her many secrets of healing; her maternal grandmother, who lives with her mother, is an expert weaver of fans. The boys in her uncle's village are admitted younger into the *Aumaga* and are not much fun when they come to call; but there are three boys in her own village whom she likes very much. And her great dilemma is whether to live with her father or her uncle, a frank, straightforward problem which introduces no ethical perplexities, no question of impersonal logic. Nor will her choice be taken as a personal matter, as the American girl's allegiance to the views of one relative might be interpreted by her other relatives. The Samoans will be sure she chose one residence rather than the other for perfectly good reasons, the food was better, she had a lover in one village, or she had quarrelled with a lover in the other village. In each case she was making concrete choices within one recognised pattern of behaviour. She was never called upon to make choices involving an actual rejection of the standards of her social group, such as the daughter of Puritan parents, who permits indiscriminate caresses, must make in our society.

And not only are our developing children faced by a series of groups advocating different and mutually exclusive standards, but a more perplexing problem presents itself to them. Because our civilisation is woven of so many diverse strands, the ideas which any one group accepts will be found to contain numerous contradictions. So if the girl has given her allegiance whole-heartedly to some one group and has accepted in good faith their asseverations that they alone are right and other philosophies of life are Antichrist and anathema, her troubles are still not over. While the less thoughtful receives her worst blows in the discovery that what father thinks is good, grandfather thinks is bad, and that things which are permitted at home are banned at school, the more thoughtful child has subtler difficulties in store for her. If she has philosophically accepted the fact that there are several standards among which she must choose, she may still preserve a child-like faith in the coherence of her chosen philosophy. Beyond the immediate choice which was so puzzling and hard to make, which perhaps involved hurting her parents or alienating her friends, she expects peace. But she has not reckoned with the fact that each of the philosophies with which she is confronted is itself but the half-ripened fruit of compromise. If she accept Christianity, she is immediately confused between the Gospel teachings concerning peace and the value of human life and the Church's whole-hearted acceptance of war. The compromise made seventeen centuries ago between the Roman philosophy of war and domination, and the early Church doctrine of peace and humility, is still present to confuse the modern child. If she accepts the philosophic premises upon which the Declaration of Independence of the United States was founded, she finds herself faced with the necessity of reconciling the belief in the equality of man and our institutional pledges of equality of opportunity with our treatment of the Negro and the Oriental. The diversity of standards in present-day society is so striking that the dullest, the most incurious, cannot fail to notice it. And this diversity is so old, so embodied in semi-solutions, in those compromises between different philosophies which we call Christianity, or democracy, or humanitarianism, that it baffles the most intelligent, the most curious, the most analytical.

So for the explanation of the lack of poignancy in the choices of growing girls in Samoa, we must look to the temperament of the Samoan civilisation which discounts strong feeling. But for the explanation of the lack of conflict we must

look principally to the difference between a simple, homogenous primitive civilisation, a civilisation which changes so
slowly that to each generation it appears static, and a motley,
diverse, heterogeneous modern civilisation.

And in making the comparison there is a third consideration, the lack of neuroses among the Samoans, the great number of neuroses among ourselves. We must examine the factors
in the early education of the Samoan children which have
fitted them for a normal, unneurotic development. The findings of the behaviourists and of the psychoanalysts alike lay
great emphasis upon the enormous rôle which is played by the
environment of the first few years. Children who have been
given a bad start are often found to function badly later on
when they are faced with important choices. And we know
that the more severe the choice, the more conflict; the more
poignancy is attached to the demands made upon the individual, the more neuroses will result. History, in the form of
the last war, provided a stupendous illustration of the great
number of maimed and handicapped individuals whose defects
showed only under very special and terrible stress. Without
the war, there is no reason to believe that many of these
shell-shocked individuals might not have gone through life
unremarked; the bad start, the fears, the complexes, the bad
conditionings of early childhood, would never have borne
positive enough fruit to attract the attention of society.

The implications of this observation are double. Samoa's
lack of difficult situations, of conflicting choice, of situations
in which fear or pain or anxiety are sharpened to a knife edge
will probably account for a large part of the absence of psychological maladjustment. Just as a low-grade moron would
not be hopelessly handicapped in Samoa, although he would
be a public charge in a large American city, so individuals
with slight nervous instability have a much more favourable
chance in Samoa than in America. Futhermore the amount
of individualisation, the range of variation, is much smaller in
Samoa. Within our wider limits of deviation there are inevitably found weak and non-resistant temperaments. And just as
our society shows a greater development of personality, so also
it shows a larger proportion of individuals who have suc-
cumbed before the complicated exactions of modern life.

Nevertheless, it is possible that there are factors in the early
environment of the Samoan child which are particularly favourable to the establishment of nervous stability. Just as a
child from a better home environment in our civilisation may
be presumed to have a better chance under all circumstances
it is conceivable that the Samoan child is not only handled

more gently by its culture but that it is also better equipped for those difficulties which it does meet.

Such an assumption is given force by the fact that little Samoan children pass apparently unharmed through experiences which often have grave effects on individual development in our civilisation. Our life histories are filled with the later difficulties which can be traced back to some early, highly charged experience with sex or with birth or death. And yet Samoan children are familiarised at an early age and without disaster, with all three. It is very possible that there are aspects of the life of the young child in Samoa which equip it particularly well for passing through life without nervous instability.

With this hypothesis in mind it is worth while to consider in more detail which parts of the young child's social environment are most strikingly different from ours. Most of these centre about the family situation, the environment which impinges earliest and most intensely upon the child's consciousness. The organisation of a Samoan household eliminates at one stroke, in almost all cases, many of the special situations which are believed to be productive of undesirable emotional sets. The youngest, the oldest, and the only child, hardly ever occur because of the large number of children in a household, all of whom receive the same treatment. Few children are weighted down with responsibility, or rendered domineering and overbearing as eldest children so often are, or isolated, condemned to the society of adults and robbed of the socialising effect of contact with other children, as only children so often are. No child is petted and spoiled until its view of its own deserts is hopelessly distorted, as is so often the fate of the youngest child. But in the few cases where Samoan family life does approximate ours, the special attitudes incident to order of birth and to close affectional ties with the parent tend to develop.

The close relationship between parent and child, which has such a decisive influence upon so many in our civilisation, that submission to the parent or defiance of the parent may become the dominating pattern of a lifetime, is not found in Samoa. Children reared in households where there are a half dozen adult women to care for them and dry their tears, and a half dozen adult males, all of whom represent constituted authority, do not distinguish their parents as sharply as our children do. The image of the fostering, loving mother, or the admirable father, which may serve to determine affectional choices later in life, is a composite affair, composed of several aunts, cousins, older sisters and grandmothers; of chief, father, uncles, brothers and cousins. Instead of learning as its first lesson that here is a kind mother whose special and principal care is for its welfare, and a father whose authority is to be deferred to, the Samoan baby learns that its world is composed of a hierarchy of male and female adults, all of whom can be depended upon and must be deferred to.

The lack of specialised feeling which results from this diffusion of affection in the household is further reinforced by the segregation of the boys from the girls, so that a child regards the children of the opposite sex as taboo relatives, regardless of individuality, or as present enemies and future lovers, again regardless of individuality. And the substitution of relationship for preference in forming friendship completes the work. By the time she reaches puberty the Samoan girl has learned to subordinate choice in the selection of friends or lovers to an observance of certain categories. Friends must be relatives of one's own sex; lovers, non-relatives. All claim of personal attraction or congeniality between relatives of opposite sex must be flouted. All of this means that casual sex relations carry no onus of strong attachment, that the marriage of convenience dictated by economic and social considerations is easily born and casually broken without strong emotion.

Nothing could present a sharper contrast to the average American home, with its small number of children, the close, theoretically permanent tie between the parents, the drama of the entrance of each new child upon the scene and the deposition of the last baby. Here the growing girl learns to depend upon a few individuals, to expect the rewards of life from certain kinds of personalities. With this first set towards preference in personal relations she grows up playing with boys as well as with girls, learning to know well brothers and cousins and school mates. She does not think of boys as a class but as individuals, nice ones like the brother of whom she is fond, or disagreeable, domineering ones, like a brother with whom she is always on bad terms. Preference in physical make-up, in temperament, in character, develops and forms the foundations for a very different adult attitude in which choice plays a vivid rôle. The Samoan girl never tastes the rewards of romantic love as we know it, nor does she suffer as an old maid who has appealed to no lover or found no lover appealing to her, or as the frustrated wife in a marriage which has not fulfilled her high demands.

Having learned a little of the art of disciplining sex feeling into special channels approved by the whole personality, we will be inclined to account our solution better than the Samoans. To attain what we consider a more dignified standard of personal relations we are willing to pay the penalty of frigidity in marriage and a huge toll of barren, unmarried women who move in unsatisfied procession across the American and English stage. But while granting the desirability of this development of sensitive, discriminating response to personality, as a better basis for dignified human lives than an automatic, undifferentiated response to sex attraction, we may still, in the light of Samoan solutions, count our methods exceedingly expensive.

The strict segregation of related boys and girls, the institutionalised hostility between pre-adolescent children of opposite sexes in Samoa are cultural features with which we are completely out of sympathy. For the vestiges of such attitudes, expressed in our one-sex schools, we are trying to substitute coeducation, to habituate one sex to another sufficiently so that difference of sex will be lost sight of in the more important and more striking differences in personality. There are no recognisable gains in the Samoan system of taboo and segregation, of response to a group rather than response to an individual. But when we contrast the other factor of difference the conclusion is not so sure. What are the rewards of the tiny, ingrown, biological family opposing its closed circle of affection to a forbidding world, of the strong ties between parents and children, ties which imply an active personal relation from birth until death? Specialisation of affection, it is true, but at the price of many individuals' preserving through life the attitudes of dependent children, of ties between parents and children which successfully defeat the children's attempts to make other adjustments, of necessary choices made unnecessarily poignant because they become issues in an intense emotional relationship. Perhaps these are too heavy prices to pay for a specialisation of emotion which might be brought about in other ways, notably through coeducation. And with such a question in our minds it is interesting to note that a larger family community, in which there are several adult men and women, seems to ensure the child against the development of the crippling attitudes which have been labelled Œdipus complexes, Electra complexes, and so on.

The Samoan picture shows that it is not necessary to channel so deeply the affection of a child for its parents and suggests that while we would reject that part of the Samoan scheme which holds no rewards for us, the segregation of the sexes before puberty, we may learn from a picture in which the home does not dominate and distort the life of the child.

The presence of many strongly held and contradictory points of view and the enormous influence of individuals in the lives of their children in our country play into each other's hands in producing situations fraught with emotion and pain. In Samoa the fact that one girl's father is a domineering, dogmatic person, her cousin's father a gentle, reasonable person, and another cousin's father a vivid, brilliant, eccentric person, will influence the three girls in only one respect, choice of residence if any one of the three fathers is the head of a household. But the attitudes of the three girls towards sex, and towards religion, will not be affected by the different temperaments of their three fathers, for the fathers play too slight a rôle in their lives. They are schooled not by an individual but by an army of relatives into a general conformity upon which the personality of their parents has a very slight effect. And through an endless chain of cause and effect, individual differences of standard are not perpetuated through the children's adherence to the parents' position, nor are children thrown into bizarre, untypical attitudes which might form the basis for departure and change. It is possible that where our own culture is so charged with choice, it would be desirable to mitigate, at least in some slight measure, the strong rôle which parents play in children's lives, and so eliminate one of the most powerful accidental factors in the choices of any individual life.

The Samoan parent would reject as unseemly and odious an ethical plea made to a child in terms of personal affection. "Be good to please mother." "Go to church for father's sake." "Don't be so disagreeable to your sister, it makes father so unhappy." Where there is one standard of conduct and only one, such undignified confusion of ethics and affection is blessedly eliminated. But where there are many standards and all adults are striving desperately to bind their own children to the particular courses which they themselves have chosen, recourse is had to devious and non-reputable means. Beliefs, practices, courses of action, are pressed upon the child in the name of filial loyalty. In our ideal picture of the freedom of the individual and the dignity of human relations it is not pleasant to realise that we have developed a form of family organisation which often cripples the emotional life, and warps and confuses the growth of many individuals' power to consciously live their own lives.

The third element in the Samoan pattern of lack of personal relationships and lack of specialised affection, is the case of friendship. Here, most of all, individuals are placed in categories and the response is to the category, "relative," or "wife of my husband's talking chief," or "son of my father's talking chief," or "daughter of my father's talking chief." Consideration of congeniality, of like-mindedness, are all ironed out in favour of regimented associations. Such attitudes we would of course reject completely.

Drawing the threads of this particular discussion together,

we may say that one striking difference between Samoan society and our own is the lack of the specialisation of feeling, and particularly of sex feeling, among the Samoans. To this difference is undoubtedly due a part of the lack of difficulty of marital adjustments in a marriage of convenience, and the lack of frigidity or psychic impotence. This lack of specialisation of feeling must be attributed to the large heterogeneous household, the segregation of the sexes before adolescence, and the regimentation of friendship—chiefly along relationship lines. And yet, although we deplore the prices in maladjusted and frustrated lives, which we must pay for the greater specialisation of sex feeling in our own society, we nevertheless vote the development of specialised response as a gain which we would not relinquish. But an examination of these three casual factors suggest that we might accomplish our desired end, the development of a consciousness of personality, through coeducation and free and unregimented friendships, and possibly do away with the evils inherent in the too intimate family organisation, thus eliminating a part of our penalty of maladjustment without sacrificing any of our dearly bought gains.

The next great difference between Samoa and our own culture which may be credited with a lower production of maladjusted individuals is the difference in the attitude towards sex and the education of the children in matters pertaining to birth and death. None of the facts of sex or of birth are regarded as unfit for children, no child has to conceal its knowledge for fear of punishment or ponder painfully over little-understood occurrences. Secrecy, ignorance, guilty knowledge, faulty speculations resulting in grotesque conceptions which may have far-reaching results, a knowledge of the bare physical facts of sex without a knowledge of the accompanying excitement, of the fact of birth without the pains of labour, of the fact of death without the fact of corruption—all the chief flaws in our fatal philosophy or sparing children a knowledge of the dreadful truth—are absent in Samoa. Furthermore, the Samoan child who participates intimately in the lives of a host of relatives has many and varied experiences upon which to base its emotional attitudes. Our children, confined within one family circle (and such confinement is becoming more and more frequent with the growth of cities and the substitution of apartment houses with a transitory population for a neighbourhood of householders), often owe their only experience with birth or death to the birth of a younger brother or sister or the death of a parent or grandparent. Their knowledge of sex, aside from children's gossip, comes from an accidental glimpse of parental activity. This has several very obvious disadvantages. In the first place, the child is dependent for its knowledge upon birth and death entering its own home; the youngest child in a family where there are no deaths may grow to adult life without ever having had any close knowledge of pregnancy, experience with young children, or contact with death.

A host of ill-digested fragmentary conceptions of life and death will fester in the ignorant, inexperienced mind and provide a fertile field for the later growth of unfortunate attitudes. Second, such children draw their experiences from too emotionally toned a field; one birth may be the only one with which they come in close contact for the first twenty years of their lives. And upon the accidental aspects of this particular birth their whole attitude is dependent. If the birth is that of a younger child who usurps the elder's place, if the mother dies in child bed, or if the child which is born is deformed, birth may seem a horrible thing, fraught with only unwelcome consequences. If the only death bed at which one has ever watched is the death bed of one's mother, the bare fact of death may carry all the emotion which that bereavement aroused, carry forever an effect out of all proportion to the particular deaths encountered later in life. And intercourse seen only once or twice, between relatives towards whom the child has complicated emotional attitudes, may produce any number of false assumptions. Our records of maladjusted children are full of cases where children have misunderstood the nature of the sexual act, have interpreted it as struggle accompanied by anger, or as chastisement, have recoiled in terror from one highly charged experience. So our children are dependent upon accident for their experience of life and death; and those experiences which they are vouchsafed, lie within the intimate family circle and so are the worst possible way of learning general facts about which it is important to acquire no special, distorted attitudes. One death, two births, one sex experience, is a generous total for the child brought up under living conditions which we consider consonant with an American standard of living. And considering the number of illustrations which we consider it necessary to give of how to calculate the number of square feet of paper necessary to paper a room eight feet by twelve feet by fourteen feet, or how to parse an English sentence, this is a low standard of illustration. It might be argued that these are experiences of such high emotional tone that repetition is unnecessary. It might also be argued if a child were severely beaten before being given its first lesson in calculating how to paper a room, and as a sequel to the lesson, saw its father hit its mother with the poker, it would always remember that arithmetic lesson. But what it would know about the real nature of the calculations involved in room-papering is doubtful. In one or two experiences, the child is given no perspective, no chance to relegate the grotesque and unfamiliar physical details of the life process to their proper place. False impressions, part impressions, repulsion, nausea, horror, grow up about some fact experienced only once under intense emotional stress and in an atmosphere unfavourable to the child's attaining any real understanding.

A standard of reticence which forbids the child any sort of comment upon its experiences makes for the continuance of such false impressions, such hampering emotional attitudes, questions such as, "Why were grandma's lips so blue?" are promptly hushed. In Samoa, where decomposition sets in almost at once, a frank, naïve repugnance to the odours of corruption on the part of all the participants at a funeral robs the physical aspect of death of any special significance. So, in our arrangements, the child is not allowed to repeat his experiences, and he is not permitted to discuss those which he has had and correct his mistakes.

With the Samoan child it is profoundly different. Intercourse, pregnancy, child birth, death, are all familiar occurrences. And the Samoan child experiences them in no such ordered fashion as we, were we to decide for widening the child's experimental field, would regard as essential. In a civilisation which suspects privacy, children of neighbours will be accidental and unemotional spectators in a house where the head of the household is dying or the wife is delivered of a miscarriage. The pathology of the life processes is known to them, as well as the normal. One impression corrects an earlier one until they are able, as adolescents, to think about life and death and emotion without undue preoccupation with the purely physical details.

It must not be supposed, however, that the mere exposure of children to scenes of birth and death would be a sufficient guarantee against the growth of undesirable attitudes. Probably even more influential than the facts which are so copiously presented to them, is the attitude of mind with which their elders regard the matter. To them, birth and sex and death are the natural, inevitable structure of existence, of an existence in which they expect their youngest children to share. Our so often repeated comment that "it's not natural" for children to be permitted to encounter death would seem as incongruous to them as if we were to say it was not natural for children to see other people eat or sleep. And this calm, matter-of-fact acceptance of their children's presence envelops the children in a protective atmosphere, saves them from shock and binds them closer to the common emotion which is so dignifiedly permitted them.

As in every case, it is here impossible to separate attitude from practice and say which is primary. The distinction is made only for our use in another civilisation. The individual American parents, who believe in a practice like the Samoan, and permit their children to see adult human bodies and gain a wider experience of the functioning of the human body than is commonly permitted in our civilisation, are building upon sand. For the child, as soon as it leaves the protecting circle of its home, is blasted by an attitude which regards such experience in children as ugly and unnatural. As likely as not, the attempt of the individual parents will have done the child more harm than good, for the necessary supporting social attitude is lacking. This is just a further example of the possibilities of maladjustment inherent in a society where each home differs from each other home; for it is in the fact of difference that the strain lies rather than in the nature of the difference.

Upon this quiet acceptance of the physical facts of life, the Samoans build, as they grow older, an acceptance of sex. Here again it is necessary to sort out which parts of their practice seem to produce results which we certainly deprecate, and which produce results which we desire. It is possible to analyse Samoan sex practice from the standpoint of development of personal relationships on the one hand, and of the obviation of specific difficulties upon the other.

We have seen that the Samoans have a low level of appreciation of personality differences, and a poverty of conception of personal relations. To such an attitude the acceptance of promiscuity undoubtedly contributes. The contemporaneousness of several experiences, their short duration, the definite avoidance of forming any affectional ties, the blithe acceptance of the dictates of a favourable occasion, as in the expectation of infidelity in any wife whose husband is long from home, all serve to make sex an end rather than a means, something which is valued in itself, and deprecated inasmuch as it tends to bind one individual to another. Whether such a disregard of personal relations is completely contingent upon the sex habits of the people is doubtful. It probably is also a reflection of a more general cultural attitude in which personality is consistently disregarded. But there is one respect in which these very practices make possible a recognition of personality which is often denied to many in our civilisation, because, from the Samoans' complete knowledge of sex, its possibilities and its rewards, they are able to count it at its true value. And if they have no preference for reserving sex activity for important relationships, neither do they regard relationships as important because they are productive of sex satisfaction. The Samoan girl who shrugs her shoulder over the excellent technique of some young Lothario is nearer to the recognition of sex as an impersonal force without any intrinsic validity, than is the sheltered American girl who falls in love with the first man who kisses her. From their familiarity with the reverberations which accompany sex excitement comes this recognition of the essential impersonality of sex attraction which we may well envy them; from the too slight, too casual practice comes the disregard of personality which seems to us unlovely.

The fashion in which their sex practice reduces the possibility of neuroses has already been discussed. By discounting our category of perversion, as applied to practice, and reserving it for the occasional psychic pervert, they legislate a whole field of neurotic possibility out of existence. Onanism, homosexuality, statistically unusual forms of heterosexual activity, are neither banned nor institutionalised. The wider range which these practices give prevents the development of obsessions of guilt which are so frequent a cause of maladjustment among us. The more varied practices permitted heterosexually preserve any individual from being penalised for special conditioning. This acceptance of a wider range as "normal" provides a cultural atmosphere in which frigidity and psychic impotence do not occur and in which a satisfactory sex adjustment in marriage can always be established. The acceptance of such an attitude without in any way accepting promiscuity would go a long way towards solving many marital impasses and emptying our park benches and our houses of prostitution.

Among the factors in the Samoan scheme of life which are influential in producing stable, well-adjusted, robust individuals, the organisation of the family and the attitude towards sex are undoubtedly the most important. But it is necessary to note also the general educational concept which disapproves of precocity and coddles the slow, the laggard, the inept. In a society where the tempo of life was faster, the rewards greater, the amount of energy expended larger, the bright children might develop symptoms of boredom. But the slower pace dictated by the climate, the complacent, peaceful society, and the compensation of the dance, in its blatant precocious display of individuality which drains off some of the discontent which the bright child feels, prevent any child from becoming too bored. And the dullard is not goaded and dragged along faster than he is is able until, sick with making an impossible effort, he gives up entirely. This educational policy also tends to blur individual differences and so to minimise jealousy, rivalry, emulation, those social attitudes which arise out of discrepancies of endowment and are so far-reaching in their effects upon the adult personality.

It is one way of solving the problem of differences between individuals and a method of solution exceedingly congenial to a strict adult world. The longer the child is kept in a subject, non-initiating state, the more of the general cultural attitude it will absorb, the less of a disturbing element it will become. Furthermore, if time is given them, the dullards can learn enough to provide a stout body of conservatives upon whose shoulders the burden of the civilisation can safely rest. Giving titles to young men would put a premium upon the exceptional; giving titles to men of forty, who have at last acquired sufficient training to hold them, assures the continuation of the usual. It also discourages the brilliant so that their social contribution is slighter than it might otherwise have been.

We are slowly feeling our way towards a solution of this problem, at least in the case of formal education. Until very recently our educational system offered only two very partial solutions of the difficulties inherent in a great discrepancy between children of different endowment and different rates of development. One solution was to allow a sufficiently long time to each educational step so that all but the mentally defective could succeed, a method similar to the Samoan one and without its compensatory dance floor. The bright child, held back, at intolerably boring tasks, unless he was fortunate enough to find some other outlet for his unused energy, was likely to expend it upon truancy and general delinquency. Our only alternative to this was "skipping," a child from one grade to another, relying upon the child's superior intelligence to bridge the gaps. This was a method congenial to American enthusiasm for meteoric careers from canal boat and log cabin to the White House. Its disadvantages in giving the child a sketchy, discontinuous background, in removing it from its age group, have been enumerated too often to need repetition here. But it is worthy of note that with a very different valuation of individual ability than that entertained by Samoan society we used for years one solution, similar and less satisfactory than theirs, in our formal educational attempts.

The methods which experimental educators are substituting for these unsatisfactory solutions, schemes like the Dalton Plan, or the rapidly moving classes in which a group of children can move ahead at a high, even rate of speed without hurt to themselves or to their duller fellows, is a striking example of the results of applying reason to the institutions of our society. The old red school-house was almost as haphazard and accidental a phenomenon as the Samoan dance floor. It was an institution which had grown up in response to a vaguely felt, unanalysed need. Its methods were analogous to the methods used by primitive peoples, non-rationalised solutions of pressing problems. But the institutionalisation of different methods of education for children of different capacities and different rates of development is not like anything which we find in Samoa or in any other primitive society. It is the conscious, intelligent directing of human institutions in response to observed human needs.

Still another factor in Samoan education which results in different attitudes is the place of work and play in the children's lives. Samoan children do not learn to work through learning to play, as the children of many primitive peoples do. Nor are they permitted a period of lack of responsibility such as our children are allowed. From the time they are four or five years old they perform definite tasks, graded to their strength and intelligence, but still tasks which have a meaning in the structure of the whole society. This does not mean that they have less time for play than American children who are shut up in schools from nine to three o'clock every day. Before the introduction of schools to complicate the ordered routine of their lives, the time spent by the Samoan child in running errands, sweeping the house, carrying water, and taking actual care of the baby, was possibly less than that which the American school child devotes to her studies.

The difference lies not in the proportion of time in which their activities are directed and the proportion of time in which they are free, but rather in the difference of attitude. With the professionalisation of education and the specialisation of industrial tasks which has stripped the individual home of its former variety of activities, our children are not made to feel that the time they do devote to supervised activity is functionally related to the world of adult activity. Although this lack of connection is more apparent than real, it is still sufficiently vivid to be a powerful determinant in the child's attitude. The Samoan girl who tends babies, carries water, sweeps the floor; or the little boy who digs for bait, or collects cocoanuts, has no such difficulty. The necessary nature of their tasks is obvious. And the practice of giving a child a task which he can do well and never permitting a childish, inefficient tinkering with adult apparatus, such as we permit to our children, who bang aimlessly and destructively on their fathers' typewriters, results in a different attitude towards work. American children spend hours in

schools learning tasks whose visible relation to their mothers'
and fathers' activities is often quite impossible to recognise.
Their participation in adults' activities is either in terms of
toys, tea-sets and dolls and toy automobiles, or else a meaningless and harmful tampering with the electric light system.
(It must be understood that here, as always, when I say
American, I do not mean those Americans recently arrived
from Europe, who still present a different tradition of education. Such a group would be the Southern Italians, who still
expect productive work from their children.)

So our children make a false set of categories, work, play,
and school; work for adults, play for children's pleasure, and
schools as an inexplicable nuisance with some compensations.
These false distinctions are likely to produce all sorts of
strange attitudes, an apathetic treatment of a school which
bears no known relation to life, a false dichotomy between
work and play, which may result either in a dread of work
as implying irksome responsibility or in a later contempt for
play as childish.

The Samoan child's dichotomy is different. Work consists
of those necessary tasks which keep the social life going:
planting and harvesting and preparation of food, fishing.
house-building, mat-making, care of children, collecting of
property to validate marriages and births and succession to
titles and to entertain strangers, these are the necessary activities of life, activities in which every member of the community, down to the smallest child, has a part. Work is not a
way of acquiring leisure; where every household produces
its own food and clothes and furniture, where there is no
large amount of fixed capital and households of high rank
are simply characterised by greater industry in the discharge
of greater obligations, our whole picture of saving, of invest-
ment, of deferred enjoyment, is completely absent. (There is
even a lack of clearly defined seasons of harvest, which would
result in special abundance of food and consequent feasting.
Food is always abundant, except in some particular village
where a few weeks of scarcity may follow a period of lavish
entertaining.) Rather, work is something which goes on all
the time for every one; no one is exempt; few are overworked.
There is social reward for the industrious, social toleration
for the man who does barely enough. And there is always
leisure—leisure, be it noted, which is not the result of hard
work or accumulated capital at all, but is merely the result
of a kindly climate, a small population, a well-integrated
social system, and no social demands for spectacular expen-
diture. And play is what one does with the time left over

from working, a way of filling in the wide spaces in a structure of unirksome work.

Play includes dancing, singing, games, weaving necklaces of flowers, flirting, repartee, all forms of sex activity. And there are social institutions like the ceremonial inter-village visit which partake of both work and play. But the distinctions between work as something one has to do but dislikes, and play as something one wants to do; of work as the main business of adults, play as the main concern of children, are conspicuously absent. Children's play is like adults' play in kind, interest, and in its proportion to work. And the Samoan child has no desire to turn adults' activities into play, to translate one sphere into the other. I had a box of white clay pipes for blowing soap bubbles sent me. The children were familiar with soap bubbles, but their native method of blowing them was very inferior to the use of clay pipes. But after a few minutes' delight in the unusual size and beauty of the soap bubbles, one little girl after another asked me if she might please take her pipe home to her mother, for pipes were meant to smoke, not to play with. Foreign dolls did not interest them, and they have no dolls of their own, although children of other islands weave dolls from the palm leaves from which Samoan children weave balls. They never make toy houses, nor play house, nor sail toy boats. Little boys would climb into a real outrigger canoe and practise paddling it within the safety of the lagoon. This whole attitude gave a greater coherence to the children's lives than we often afford our children.

The intelligibility of a child's life among us is measured only in terms of the behaviour of other children. If all the other children go to school the child who does not feels incongruous in their midst. If the little girl next door is taking music lessons, why can't Mary; or why must Mary take music lessons, if the other little girl doesn't take them. But so sharp is our sense of difference between the concerns of children and of adults that the child does not learn to judge its own behaviour in relationship to adult life. So children often learn to regard play as something inherently undignified, and as adults mangle pitifully their few moments of leisure. But the Samoan child measures her every act of work or play in terms of her whole community; each item of conduct is dignified in terms of its realised relationship to the only standard she knows, the life of a Samoan village. So complex and stratified a society as ours cannot hope to develop spontaneously any such simple scheme of education. Again we will be hard put to it to devise ways of participation for children, and means of articulating their school life with the rest of life which will give them the same dignity which Samoa affords her children.

Last among the cultural differences which may influence the emotional stability of the child is the lack of pressure to make important choices. Children are urged to learn, urged to behave, urged to work, but they are not urged to hasten in the choices which they make themselves. The first point at which this attitude makes itself felt is in the matter of the brother and sister taboo, a cardinal point of modesty and decency. Yet the exact stage at which the taboo should be observed is always left to the younger child. When it reaches a point of discretion, of understanding, it will of itself feel "ashamed" and establish the formal barrier which will last until old age. Likewise, sex activity is never urged upon the young people, nor marriage forced upon them at a tender age. Where the possibilities of deviation from the accepted standard are so slight, a few years leeway holds no threat for the society. The child who comes later to a realisation of the brother and sister taboo really endangers nothing.

This laissez-faire attitude has been carried over into the Samoan Christian Church. The Samoan saw no reason why young unmarried people should be pressed to make momentous decisions which would spoil part of their fun in life. Time enough for such serious matters after they were married or later still, when they were quite sure of what steps they were taking and were in less danger of falling from grace every month or so. The missionary authorities, realizing the virtues of going slowly and sorely vexed to reconcile Samoan sex ethics with a Western European code, saw the great disadvantages of unmarried Church members who were not locked up in Church schools. Consequently, far from urging the adolescent to think upon her soul the native pastor advises her to wait until she is older, which she is only too glad to do.

But, especially in the case of our Protestant churches, there is a strong preference among us for the appeal to youth. The Reformation, with its emphasis upon individual choice, was unwilling to accept the tacit habitual Church membership which was the Catholic pattern, a membership marked by additional sacramental gifts but demanding no sudden conversion, no renewal of religious feeling. But the Protestant solution is to defer the choice only so far as necessary, and the moment the child reaches an age which may be called "years of discretion" it makes a strong, dramatic appeal. This appeal is reinforced by parental and social pressure; the child is bidden to choose now and wisely. While such a position in the churches which stem from the Reformation and its strong emphasis on individual choice was historically inevitable, it is regrettable that the convention has lasted so long. It has even been taken over by non-sectarian reform groups, all of whom regard the adolescent child as the most legitimate field of activity.

In all of these comparisons between Samoan and American culture, many points are useful only in throwing a spotlight upon our own solutions, while in others it is possible to find suggestions for change. Whether or not we envy other peoples one of their solutions, our attitude towards our own solutions must be greatly broadened and deepened by a consideration of the way in which other peoples have met the same problems. Realising that our own ways are not humanly inevitable nor God-ordained, but are the fruit of a long and turbulent history, we may well examine in turn all of our institutions, thrown into strong relief against the history of other civilisations, and weighing them in the balance, be not afraid to find them wanting.

14

Education for Choice

WE HAVE BEEN comparing point for point, our civilisation and the simpler civilisation of Samoa, in order to illuminate our own methods of education. If now we turn from the Samoan picture and take away only the main lesson which we learned there, that adolescence is not necessarily a time of stress and strain, but that cultural conditions make it so, can we draw any conclusions which might bear fruit in the training of our adolescents?

At first blush the answer seems simple enough. If adolescents are only plunged into difficulties and distress because of conditions in their social environment, then by all means let us so modify that environment as to reduce this stress and eliminate this strain and anguish of adjustment. But, unfortunately, the conditions which vex our adolescents are the flesh and bone of our society, no more subject to straightforward manipulation upon our part than is the language which we speak. We can alter a syllable here, a construction there; but the great and far-reaching changes in linguistic structure (as in all parts of culture) are the work of time, a work in which each individual plays an unconscious and inconsiderable part. The principal causes of our adolescents' difficulty are the presence of conflicting standards and the belief that every individual should make his or her own choices, coupled with a feeling that choice is an important matter. Given these cultural attitudes, adolescence, regarded now not as a period of physiological change, for we know that physiological puberty need not produce conflict, but as the beginning of mental and emotional maturity, is bound to be filled with conflicts and difficulties. A society which is clamouring for choice, which is filled with many articulate groups, each urging its own brand of salvation, its own variety of economic philosophy, will give each new generation no peace until all have chosen or gone under, unable to bear the conditions of choice. The stress is in our civilisation, not in the physical changes through which our children pass, but it is none the less real nor less inevitable in twentieth-century America.

And if we look at the particular forms which this need for choice takes, the difficulty of the adolescent's position is only documented further. Because the discussion is principally concerned with girls, I shall discuss the problem from the girls' point of view, but in many respects the plight of the adolescent boy is very similar. Between fourteen and eighteen, the average American boy and girl finish school. They are now ready to go to work and must choose what type of work they wish to do. It might be argued that they often have remarkably little choice. Their education, the part of the country in which they live, their skill with their hands, will combine to dictate choice perhaps between the job of cash girl in a department store or of telephone operator, or of clerk or miner. But small as is the number of choices open to them in actuality, the significance of this narrow field of opportunity is blurred by our American theory of endless possibilities. Moving picture, magazine, newspaper, all reiterate the Cinderella story in one form of another, and often the interest lies as much in the way cash girl 456 becomes head buyer as in her subsequent nuptials with the owner of the store. Our occupational classes are not fixed. So many children are better educated and hold more skilled positions than their parents that even the ever-present discrepancy between opportunities open to men and opportunities open to women, although present in a girl's competition with her brother, is often absent as between her unskilled father and herself. It is needless to argue that these attitudes are products of conditions which no longer exist, particularly the presence of a frontier and a large amount of free land which provided a perpetual alternative of occupational choice. A set which was given to our thinking in pioneer days is preserved in other terms. As long as we have immigrants from non-English-speaking countries, the gap in opportunities between non-English-speaking parents and English-speaking children will be vivid and dramatic. Until our standard of education becomes far more stable than it is present, the continual raising of the age and grade until which schooling is compulsory ensures a wide educational gap between many parents and their children. And occupational shifts like the present movements of farmers and farm workers into urban occupations, give the same picture. When the agricultural worker pictures urban work as a step up in the social scale, and the introduction of scientific farming is so radically reducing the numbers needed in agriculture, the movement of young people born on the farm to city jobs is bound to dazzle the imagination of our farming states during the next generation at least. The substitution of machines for unskilled workers and the absorption of many of the workers and their children into positions where they manipulate machines affords another instance of the kind of historical change which keeps our myth of endless opportunity alive. Add to these special features, like the effect upon the prospects of Negro children of the tremendous exodus from the southern corn fields, or upon the children of New England mill-hands who are deprived of an opportunity to follow dully in their parents' footsteps and must at least seek new fields if not better ones.

Careful students of the facts may tell us that class lines are becoming fixed; that while the children of immigrants make advances beyond their parents, they move up in step; that there are fewer spectacular successes among them than there used to be; that it is much more possible to predict the future status of the child from the present status of the parent. But this measured comment of the statistician has not filtered into our literature, nor our moving pictures, nor in any way served to minimise the vividness of the improvement in the children's condition as compared with the condition of their parents. Especially in cities, there is no such obvious demonstration of the fact that improvement is the rule for the children of a given class or district, and not merely a case of John Riley's making twenty dollars a week as a crossing man while Mary, his daughter, who has gone to business school, makes twenty-five dollars a week, working shorter hours. The lure of correspondence school advertising, the efflorescence of a doctrine of short-cuts to fame, all contrive to make an American boy or girl's choice of a job different from that of English children, born into a society where stratification is so old, so institutionalised, that the dullest cannot doubt it. So economic conditions force them to go to work and everything combines to make that choice a difficult one, whether in terms of abandoning a care-free existence for a confining, uncongenial one, or in terms of bitter rebellion against the choice which they must make in contrast to the opportunities which they are told are open to all Americans.

And taking a job introduces other factors of difficulty into the adolescent girl's home situation. Her dependence has always been demonstrated in terms of limits and curbs set upon her spontaneous activity in every field from spending money to standards of dress and behaviour. Because of the essentially pecuniary nature of our society, the relationship of limitation in terms of allowance to limitation of behaviour are more far-reaching than in earlier times. Parental disapproval of extreme styles of clothing would formerly have expressed itself in a mother's making her daughter's dresses high in the neck and long in the sleeve. Now it expresses itself in control through money. If Mary doesn't stop purchasing chiffon stockings, Mary shall have no money to buy stockings. Similarly, a taste for cigarettes and liquor can only be gratified through money; going to the movies, buying books and magazines of which her parents disapprove, are all dependent upon a girl's having the money, as well as upon her eluding more direct forms of control. And the importance of a supply of money in gratifying all of a girl's desires for clothes and for amusement makes money the easiest channel through which to exert parental authority. So easy is it, that the threat of cutting off an allowance, taking away the money for the one movie a week or the coveted hat, has taken the place of the whippings and bread-and-water exiles which were favourite disciplinary methods in the last century. The parents come to rely upon this method of control. The daughters come to see all censoring of their behaviour, moral, religious or social, the ethical code and the slightest sumptuary provisions in terms of an economic threat. And then at sixteen or seventeen the daughter gets a job. No matter how conscientiously she may contribute her share to the expenses of the household, it is probably only in homes where a European tradition still lingers that the wage-earning daughter gives all of her earning to her parents. (This, or course, excludes the cases where the daughter supports her parents, where the vesting of the economic responsibility in her hands changes the picture of parental control in another fashion.) For the first time in her life, she has an income of her own, with no strings of morals or of manners attached to its use. Her parents' chief instrument of discipline is shattered at one blow, but not their desire to direct their daughters' lives. They have not pictured their exercise of control as the right of those who provide, to control those for whom they provide. They have pictured it in far more traditional terms, the right of parents to control their children, an attitude reinforced by years of practising such control.

But the daughter is in the position of one who has yielded unwillingly to some one who held a whip in his hand, and now sees the whip broken. Her unwillingness to obey, her chafing under special parental restrictions which children accept as inevitable in simpler cultures, is again a feature of our conglomerate civilisation. When all the children in the community go to bed at curfew, one child is not as likely to rail against her parents for enforcing the rule. But when the little girl next door is allowed to stay up until eleven o'clock, why must Mary go to bed at eight? If all of her companions at school are allowed to smoke, why can't she? And conversely, for it is a question of the absence of a common standard far more than of the nature of the standards, if all the other little girls are given lovely fussy dresses and hats with flowers and ribbons, why must she be dressed in sensible, straight linen dress and simple round hats? Barring an excessive and passionate devotion of the children to their parents, a devotion of a type which brings other more serious difficulties in its wake, children in a heterogeneous civilisation will not accept unquestioningly their parents' judgment, and the most obedient will temper present compliance with the hope of future emancipation.

In a primitive, homogenous community, disciplinary measures of parents are expended upon securing small concessions from children, in correcting the slight deviations which occur within one pattern of behaviour. But in our society, home discipline is used to establish one set of standards as over against other sets of standards, each family group is fighting some kind of battle, bearing the onus of those who follow a middle course, stoutly defending a cause already lost in the community at large, or valiantly attempting to plant a new standard far in advance of their neighbours. This propagandist aspect greatly increases the importance of home discipline in the development of a girl's personality. So we have the picture of parents, shorn of their economic authority, trying to coerce the girl who still lives beneath their roof into an acceptance of standards against which she is rebelling. In this attempt they often find themselves powerless and as a result the control of the home breaks down suddenly, and breaks down just at the point where the girl, faced with other important choices, needs a steadying home environment.

It is at about this time that sex begins to play a rôle in the girl's life, and here also conflicting choices are presented to her. If she chooses the freer standards of her own generation, she comes in conflict with her parents, and perhaps more importantly with the ideals which her parents have instilled. The present problem of the sex experimentation of young people would be greatly simplified if it were conceived of as experimentation instead of as rebellion, if no Puritan self-accusations vexed their consciences. The introduction of an experimentation so much wider and more dangerous presents sufficient problems in our lack of social canons for such behaviour. For a new departure in the field of personal relations is always accompanied by the failure of those who are not strong enough to face an unpatterned situation. Canons of honour, of personal obligation, of the limits of responsibilities, grow up only slowly. And, of first experimenters, many perish in uncharted seas. But when there is added to the pitfalls of experiment, the suspicion that the experiment is wrong and the need for secrecy, lying, fear, the strain is so great that frequent downfall is inevitable.

And if the girl chooses the other course, decides to remain true to the tradition of the last generation, she wins the sympathy and support of her parents at the expense of the comradeship of her contemporaries. Whichever way the die falls, the choice is attended by mental anguish. Only occasional children escape by various sorts of luck, a large enough group who have the same standards so that they are supported either against their parents or against the majority of their age mates, or by absorption in some other interest. But, with the exception of students for whom the problem of personal relations is sometimes mercifully deferred for a later settlement, those who find some other interest so satisfying that they take no interest in the other sex, often find themselves old maids without any opportunity to recoup their positions. The fear of spinsterhood is a fear which shadows the life of no primitive woman; it is another item of maladjustment which our civilisation has produced.

To the problem of present conduct are added all the perplexities introduced by varying concepts of marriage, the conflict between deferring marriage until a competence is assured, or marrying and sharing the expenses of the home with a struggling young husband. The knowledge of birth control, while greatly dignifying human life by introducing the element of choice at the point where human beings have before been most abjectly subject to nature, introduces further perplexities. It complicates the issue from a straight marriage-home-and-children plan of life versus independent spinsterhood by permitting marriages without children, earlier marriages, marriages and careers, sex relations without marriage and the responsibility of a home. And because the majority of girls still wish to marry and regard their occupations as stop-gaps, these problems not only influence their attitude towards men, but also their attitude towards their work, and prevent them from having a sustained interest in the work which they are forced to do.

Then we must add to the difficulties inherent in a new economic status and the necessity of adopting some standard of sex relations, ethical and religious issues to be solved. Here again the home is a powerful factor; the parents use every ounce of emotional pressure to enlist their children in one of the dozen armies of salvation. The stress of the revival meeting, the pressure of pastor and parent gives them no peace. And the basic difficulties of reconciling the teachings of authority with the practices of society and the findings of science, all trouble and perplex children already harassed beyond endurance.

Granting that society presents too many problems to her adolescents, demands too many momentous decisions on a few months' notice, what is to be done about it? One panacea suggested would be to postpone at least some of the decisions, keep the child economically dependent, or segregate her from all contact with the other sex, present her with only one set of religious ideas until she is older, more poised, better able to deal critically with the problems which will confront her. In a less articulate fashion, such an idea is back of various schemes for the prolongation of youth, through raising the working age, raising the school age, shielding school children from a knowledge of controversies like evolution versus fundamentalism, or any knowledge of sex hygiene or birth control. Even if such measures, specially initiated and legislatively enforced, could accomplish the end which they seek and postpone the period of choice, it is doubtful whether such a development would be desirable. It is unfair that very young children should be the battleground for conflicting standards, that their development should be hampered by propagandist attempts to enlist and condition them too young. It is probably equally unfair to culturally defer the decisions too late. Loss of one's fundamental religious faith is more of a wrench at thirty than at fifteen simply in terms of the number of years of acceptance which have accompanied the belief. A sudden knowledge of hitherto unsuspected aspects of sex, or a shattering of all the old conventions concerning sex behaviour, is more difficult just in terms of the strength of the old attitudes. Furthermore, in practical terms, such schemes would be as they are now, merely local, one state legislating against evolution, another against birth control, or one religious group segregating its unmarried girls. And these special local movements would simply unfit groups of young people for competing happily with children who had been permitted to make their choices earlier. Such an educational scheme, in addition to being almost impossible of execution, would be a step backward and would only beg the question.

Instead, we must turn all of our educational efforts to training our children for the choices which will confront them. Education, in the home even more than at school, instead of being a special pleading for one régime, a desperate attempt to form one particular habit of mind which will withstand all outside influences, must be a preparation for those very influences. Such an education must give far more attention to mental and physical hygiene than it has given hitherto. The child who is to choose wisely must be healthy in mind and body, handicapped in no preventable fashion. And even more importantly, this child of the future must have an open mind. The home must cease to plead an ethical cause or a religious belief with smiles or frowns, caresses or threats. The children must be taught how to think, not what to think. And because old errors die slowly, they must be taught tolerance, just as to-day they are taught intolerance. They must be taught that many ways are open to them, no one sanctioned above its alternative, and that upon them and upon them alone lies the burden of choice. Unhampered by prejudices, unvexed by too early conditioning to any one standard, they must come clear-eyed to the choices which lie before them.

For it must be realised by any student of civilisation that we pay heavily for our heterogeneous, rapidly changing civilisation; we pay in high proportions of crime and delinquency, we pay in the conflicts of youth, we pay in an ever-increasing number of neuroses, we pay in the lack of a coherent tradition without which the development of art is sadly handicapped. In such a list of prices, we must count our gains carefully, not to be discouraged. And chief among our gains must be reckoned this possibility of choice, the recognition of many possible ways of life, where other civilisations have recognized only one. Where other civilisations give a satisfactory outlet to only one temperamental type, be he mystic or soldier, business man or artist, a civilisation in which there are many standards offers a possibility of satisfactory adjustment to individuals of many different temperamental types, of diverse gifts and varying interests.

At the present time we live in a period of transition. We have many standards but we still believe that only one standard can be the right one. We present to our children the picture of a battle-field where each group is fully armoured in a conviction of the righteousness of its cause. And each of these groups make forays among the next generation. But it is unthinkable that a final recognition of the great number of ways in which man, during the course of history and at the present time, is solving the problems of life, should not bring with it in turn the downfall of our belief in a single standard. And when no one group claims ethical sanction for its customs, and each group welcomes to its midst only those who are temperamentally fitted for membership, then we shall have realised the high point of individual choice and universal toleration which a heterogeneous culture and a heterogeneous culture alone can attain. Samoa knows but one way of life and teaches it to her children. Will we, who have the knowledge of many ways, leave our children free to choose among them?

APPENDIX I

CHAPTER 4

Pages 31 to 42.

In the Samoan classification of relatives two principles, sex and age, are of the most primary importance. Relationship terms are never used as terms of address, a name or nickname being used even to father or mother. Relatives of the same age or within a year or two younger to five or ten years older are classified as of the speaker's generation, and of the same sex or of the opposite sex. Thus a girl will call her sister, her aunt, her niece, and her female cousin who are nearly of the same age, *uso,* and a boy will do the same for his brother, uncle, nephew, or male cousin. For relationships between siblings of opposite sex there are two terms, *tuafafine* and *tuagane,* female relative of the same age group of a male, and male relative of the same age group of a female. (The term *uso* has no such subdivisions.)

The next most important term is applied to younger relatives of either sex, the word *tei.* Whether a child is so classified by an older relative depends not so much on how many years younger the child may be, but rather on the amount of care that the elder has taken of it. So a girl will call a cousin two years younger than herself her *tei,* if she has lived near by, but an equally youthful cousin who has grown up in a distant village until both are grown will be called *uso.* It is notable that there is no term for elder relatives. The terms *uso, tuafafine* and *tuagane* all carry the feeling of contemporaneousness, and if it is necessary to specify seniority, a qualifying adjective must be used.

Tamā, the term for father, is applied also to the *matai* of a household, to an uncle or older cousin with whose authority a younger person comes into frequent contact and also to a much older brother who in feeling is classed with the parent generation. *Tinā* is used only a little less loosely for the mother, aunts resident in the household, the wife of the *matai* and only very occasionally for an older sister.

A distinction is also made in terminology between men's terms and women's terms for the children. A woman will say *tama* (modified by the addition of the suffixes *tane* and *fafine,* male and female) and a man will say *atalii,* son and *afafine,* daughter. Thus a woman will say, "Losa is my *tama,*" specifying her sex only when necessary. But Losa's father will speak of Losa as his *afafine.* The same usage is followed in speaking *to* a man or *to* a woman of a child. All of these terms are further modified by the addition of the word, *moni,* real, when a blood sister or blood father or mother is meant. The elders of the household are called roughly *matua,* and a grandparent is usually referred to as the *toa'ina,* the "old man" or *olamatua,* "old woman," adding an explanatory clause if necessary. All other relatives are described by the use of relative clauses, "the sister of the husband of the sister of my mother," "the brother of the wife of my brother," etc. There are no special terms for the in-law group.

CHAPTER 5

NEIGHBOURHOOD MAPS

Pages 42 to 50.

For the sake of convenience the households were numbered in sequence from one end of each village to the other. The houses did not stretch in a straight line along the beach, but were located so unevenly that occasionally one house was directly behind another. A schematic linear representation will, however, be sufficient to show the effect of location in the formation of neighbourhood groups.

VILLAGE I
Lumā

(The name of the girl will be placed under the number of the household. Adolescent girls' names in capitals, girls just reaching puberty in lower case and the pre-adolescent children in italics.)

1	2	3	4	5	6	7	8	9
	Vala		LITA	Maliu	*Lusi*	Fitu	Lia	*Fiva*
				Pola		Ula		LUNA

10	11	12	13	14	15	16	17	18
				LOTA	PALA		*Tuna*	
					Vi			
					Pele			

19	20	21	22	23	24	25	26	27
		LOSA			TULIPA	MASINA	*Mina*	Tina
							SONA	

28	29	30	31	32	33
TITA	ASO	Selu			
Sina	*Suna*	Tolo			
Elisa					

VILLAGE II
Siufaga

(Household 38 in Siufaga is adjacent to household 1 in Lumā. The two villages are geographically continuous but socially they are separate units.)

1	2	3	4	5	6	7	8	9	10
Vina		NAMU		LITA*		*Tulima*			
TOLO		TOLU							
		Lusina							

11	12	13	14	15	16	17	18	19	20
	Tatala			Lilina	Tino	MALA		LOLA *	

* Girls to whom a change of residence made important differences, see Chap. XI, "The Girl in Conflict."

21	22	23	24	25	26	27	28	29	30
Pulona	Ipu	Tasi			Tua				*Timu*
									Meta

31	32	33	34	35	36	37	38	
Lua	*Simina*						Fala	
							Solata	

VILLAGE III

Faleasao

(Faleasao was separated from Lumā by a high cliff which jutted out into the sea and made it necessary to take an inland trail to get from one seaside village to the other. This was about a twenty-minute walk from Taū. Faleasao children were looked upon with much greater hostility and suspicion than that which the children of Lumā and Siufga showed to each other. The pre-adolescent children from this village are not discussed by name and will be indicated by an *x*.)

1	2	3	4	5	6	7	8	9	10
	x	*x*	*x*	Talo	Ela	Leta			
		x	*x*						

11	12	13	14	15	16	17	18	19	20
x	*x*		Mina		Moana	Sala		*x*	Mata
x								*x*	*x*
								Luina	

21	22	23	24	25	26	27	28	29
x					*x*			*x*

CHAPTER 9

Pages 76 to 81.

The first person singular of the verb "to know," used in the negative, has two forms:

	Ta	ilo	(Contraction of Ta	te	lē	iloa)
			I	euphonic	neg.	know
				particle		

and

	Ua	le	iloa	a'u
	Pres.	neg.	know	I
	Part.			

148

The former of these expressions has a very different meaning from the latter although linguistically they represent optional syntactic forms, the second being literally, "I do not know," while the first can best be rendered by the slang phrase, "Search me." This "search me" carries no implication of lack of actual knowledge or information about the subject in question but is merely an indication either of lack of interest or unwillingness to explain. That the Samoans feel this distinction very clearly is shown by the frequent use of both forms in the same sentence: *Ta ilo ua lē iloa a'u.* "Search me, I don't know."

SAMPLE CHARACTER SKETCHES GIVEN OF MEMBERS OF THEIR HOUSEHOLDS BY ADOLESCENT GIRLS
(Literal translations from dictated texts)

I

He is an untitled man. He works hard on the plantation. He is tall, thin and dark-skinned. He is not easily angered. He goes to work and comes again at night. He is a policeman. He does work for the government. He is not filled with unwillingness. He is attractive looking. He is not married.

II

She is an old woman. She is very old. She is weak. She is not able to work. She can only remain in the house. Her hair is black. She is fat. She has elephantiasis in one leg. She has no teeth. She is not irritable. She does not hate. She is clever at weaving mats, fishing baskets and food trays.

III

She is strong and able to work. She goes inland. She weeds and makes the oven and picks breadfruit and gathers paper mulberry bark. She is kind. She is of good conduct. She is clever at weaving baskets and mats and fine mats and food trays, and painting tapa cloth and scraping and pounding and pasting paper mulberry bark. She is short, black-haired and dark-skinned. She is fat. She is good. If any one passes by she is kindly disposed towards them and calls ous, "Po'o fea 'e te maliu i ai?" (a most courteous way of asking, "Where are you going?")

IV

She is fat. She has long hair. She is dark-skinned. She is blind in one eye. She is of good behaviour. She is clever at weeding taro and weaving floor mats and fine mats. She is short. She has borne children. There is a baby. She remains in the house on some days and on other days she goes inland. She also knows how to weave baskets.

V

He is a boy. His skin is dark. So is his hair. He goes to the bush to work. He works on the taro plantation. He likes every one. He is clever at weaving baskets. He sings in the choir of the young men on Sunday. He likes very much to consort with the girls. He was expelled from the pastor's house.

VI
Portrait of herself

I am a girl. I am short. I have long hair. I love my sisters and all the people. I know how to weave baskets and fishing baskets and how to prepare paper mulberry bark. I live in the house of the pastor.

VII

He is a man. He is strong. He goes inland and works upon the plantation of his relatives. He goes fishing. He goes to gather cocoanuts and breadfruit and cooking leaves and makes the oven. He is tall. He is dark-skinned. He is rather fat. His hair is short. He is clever at weaving baskets. He braids the palm leaf thatching mats for the house.* He is also clever at house-building. He is of good conduct and has a loving countenance.

VIII

She is a woman. She can't work hard enough (to suit herself). She is also clever at weaving baskets and fine mats and at bark cloth making. She also makes the ovens and clears away the rubbish around the house. She keeps her house in fine condition. She makes the fire. She smokes. She goes fishing and gets octopuses and *tu'itu'i* (sea eggs) and comes back and eats them raw. She is kind-hearted and of loving countenance. She is never angry. She also loves her children.

IX

She is a woman. She has a son,————is his name. She is lazy. She is tall. She is thin. Her hair is long. She is clever at weaving baskets, making bark cloth and weaving fine mats. Her husband is dead. She does not laugh often. She stays in the house some days and other days she goes inland. She keeps everything clean. She lives well upon bananas. She has a loving face. She is not easily out of temper. She makes the oven.

X

She is the daughter of ————. She is a little girl about my age. She is also clever at weaving baskets and mats and fine mats and blinds and floor mats. She is good in school. She also goes to get leaves and breadfruit. She also goes fishing when the tide is out. She gets crabs and jelly fish. She is very loving. She does not eat up all her food if others ask her for it. She shows a loving face to all who come to her house. She also spreads food for all visitors.

XI
Portrait of herself

I am clever at weaving mats and fine mats and baskets and blinds and floor mats. I go and carry water for all of my household to drink and for others also. I go and gather bananas and breadfruit and

*Women's work.

leaves and make the oven with my sisters. Then we (herself and her sisters) go fishing together and then it is night.

CHAPTER 10

Pages 81 to 96.

The children of this age already show a very curious example of a phonetic self-consciousness in which they are almost as acute and discriminating as their elders. When the missionaries reduced the language to writing, there was no *k* in the language, the *k* positions in other Polynesian dialects being filled in Samoan either with a *t* or a glottal stop. Soon after the printing of the Bible, and the standardisation of Samoan spelling, greater contact wih Tonga introduced the *k* into the spoken language of Savai'i and Upolu, displacing the *t*, but not replacing the glottal stop. Slowly this intrusive usage spread eastward over Samoa, the missionaries who controlled the schools and the printing press fighting a dogged and losing battle with the less musical *k*. To-day the *t* is the sound used in the speech of the educated and in the church, still conventionally retained in all spelling and used in speeches and on occasions demanding formality. The Manu'a children who had never been to the missionary boarding schools, used the *k* entirely. But they had heard the *t* in church and at school and were sufficiently conscious of the difference to rebuke me immediately if I slipped into the colloquial *k,* which was their only speech habit, uttering the *t* sound for perhaps the first time in their lives to illustrate the correct pronunciation from which I, who was ostensibly learning to speak correctly, must not deviate. Such as ability to disassociate the sound used from the sound heard is remarkable in such very young children and indeed remarkable in any person who is not linguistically sophisticated.

CHAPTER 11

Pages 96 to 110.

During six months I saw six girls leave the pastor's establishment for several reasons: Tasi, because her mother was ill and she, that rare phenomenon, the eldest in a biological household, was needed at home; Tua, because she had come out lowest in the missionaries' annual examination which her mother attributed to favouritism on the part of the pastor; Luna, because her stepmother, whom she disliked, left her father and thus made her home more attractive and because under the influence of a promiscuous older cousin she began to tire of the society of younger girls and take an interest in love affairs; Lita, because her father ordered her home, because with the permission of the pastor, but without consulting her family, she went off for a three weeks' visit in another island. Going home for Lita involved residence in the far end of the other village, necessitating a complete change of friends. The novelty of the new group and new interests kept her from in any way chafing at the change. Sala, a stupid idle girl, had eloped from the household of the pastor.

APPENDIX II

It is impossible to present a single and unified picture of the adolescent girl in Samoa and at the same time to answer most satisfactorily the various kinds of questions which such a study will be expected to answer. For the ethnologist in search of data upon the usages and rites connected with adolescence it is necessary to include descriptions of customs which have fallen into partial decay under the impact of western propaganda and foreign example. Traditional observances and attitudes are also important in the study of the adolescent girl in present-day Samoa because they still form a large part of the thought pattern of her parents, even if they are no longer given concrete expression in the girl's cultural life. But this double necessity of describing not only the present environment and the girl's reaction to it, but also of interpolating occasionally some description of the more rigid cultural milieu of her mother's girlhood, mars to some extent the unity of the study.

The detailed observations were all made upon a group of girls living in three practically contiguous villages on one coast of the island of Taū. The data upon the ceremonial usages surrounding birth, adolescence and marriage were gathered from all of the seven villages in the Manu'a Archipelago.

The method of approach is based upon the assumption that a detailed intensive investigation will be of more value than a more diffused and general study based upon a less accurate knowledge of a greater number of individuals. Dr. Van Waters' study of *The Adolescent Girl Among Primitive Peoples* has exhausted the possibilities of an investigation based upon the merely external observations of the ethnologist who is giving a standardised description of a primitive culture. We have a huge mass of general descriptive material without the detailed observations and the individual cases in the light of which it would be possible to interpret it.

The writer therefore chose to work in one small locality, in a group numbering only six hundred people, and spent six months accumulating an intimate and detailed knowledge of all the adolescent girls in this community. As there were only sixty-eight girls between the ages of nine and twenty, quantitative statements are practically valueless for obvious reasons: the probable error of the group is too large; the age classes are too small, etc. The only point at which quantitative statements can have any relevance is in regard to the variability within the group, as the smaller the variability within the sample, the greater the general validity of the results.

Furthermore, the type of data which we needed is not of the sort which lends itself readily to quantitative treatment. The reaction of the girl to her stepmother, to relatives acting as foster parents, to her younger sister, or to her older brother,—these are incommensurable in quantitative terms. As the physician and the psychiatrist have found it necessary to describe each case separately and to use their cases as illumination of a thesis rather than as irrefutable proof such as it is possible to adduce in the physical sciences, so the student of the more intangible and psychological aspects of human behaviour is forced to illuminate rather than demonstrate a thesis. The composition of the background against which the girl acts can be described in accurate and general terms, but her reactions are a function of her own personality and cannot be described without reference to it. The generalisations are based upon a careful and detailed observation of a small group of subjects. These results will be illuminated and illustrated by case histories.

The conclusions are also all subject to the limitation of the personal equation. They are the judgments of one individual upon a mass of data, many of the most significant aspects of which can, by their very nature, be known only to herself. This was inevitable and it can only be claimed in extenuation that as the personal equation was held absolutely constant, the different parts of the data are strictly commensurable. The judgment on the reaction of Lola to her uncle and of Sona to her cousin are made on exactly the same basis.

Another methodological device which possibly needs explanation is the substitution of a cross sectional study for a linear one. Twenty-eight children who as yet showed no signs of puberty, fourteen children who would probably mature within the next year or year and a half, and twenty-five girls who had passed puberty within the last four years but were not yet classed by the community as adults, were studied in detail. Less intensive observations were also made upon the very little children and the young married women. This method of taking cross sections, samples of individuals at different periods of physical development, and arguing that a group in an earlier stage will later show the characteristics which appear in another group at a later stage, is, of course inferior to a linear study in which the same group is under observation for a number of years. A very large number of cases has usually been the only acceptable defence of such a procedure. The number of cases included in this investigation, while very small in comparison with the numbers mustered by any student of American children, is nevertheless a fair-sized sample in terms of the very small population of Samoa (a rough eight thousand in all four islands of American Samoa) and because the only selection was geographical. It may further be argued that the almost drastic character of the conclusions, the exceedingly few exceptions which need to be made, further validate the size of the sample. The adoption of the cross section method was, of course, a matter of expediency, but the results when carefully derived from a fair sample, may be fairly compared with the results obtained by using the linear method, when the same subjects are under observation over a period of years. This is true when the conclusions to be drawn are general and not individual. For the purposes of psychological theory, it is sufficient to know that children in a certain society walk, on the average, at twelve months, and talk, on the average, at fifteen months. For the purposes of the diagnostician, it is necessary to know that John walked at eighteen months and did not talk until twenty months. So, for general theoretical purposes, it is enough to state that little girls just past puberty develop a shyness and lack of self-possession in the presence of boys, but if we are to understand the delinquency of Mala, it is necessary to know that she prefers the company of boys to that of girls and has done so for several years.

Particular Methods Used

The description of the cultural background was obtained in orthodox fashion, first through interviews with carefully chosen informants, followed by checking up their statement with other informants and by the use of many examples and test cases. With a few unimportant exceptions this material was obtained in the Samoan language and not through the medium of interpreters. All of the work with individuals was done in the native language, as there were no young people on the island who spoke English.

Although a knowledge of the entire culture was essential for the accurate evaluation of any particular individual's behaviour, a detailed description will be given only of those aspects of the culture which are immediately relevant to the problem of the adolescent girl. For example, if I observe Pele refuse point blank to carry a message to the house of a relative, it is important to know whether she is actuated by stubbornness, dislike of the relative, fear of the dark, or fear of the ghost which lives near by and has a habit of jumping on people's backs. But to the reader a detailed exposition of the names and habits of all the local ghost population would be of little assistance in the appreciation of the main problem. So all descriptions of the culture which are not immediately relevant are omitted from the discussion but were not omitted from the original investigation. Their irrelevancy has, therefore, been definitely ascertained.

The knowledge of the general cultural pattern was supplemented by a detailed study of the social structure of the three villages under consideration. Each household was analysed from the standpoint of rank, wealth, location, contiguity to other households, relationship to other households, and the age, sex, relationship, marital status, number of children, former residence, etc., of each individual in the household. This material furnished a general descriptive basis for a further and more careful analysis of the households of the subjects, and also provided a check on the origin of feuds or alliances between individuals, the use of relationship terms, etc. Each child was thus studied against a background which was known in detail.

A further mass of detailed information was obtained about the subjects: approximate age (actual age can never be determined in Samoa), order of birth, numbers of brothers and sisters, who were older and younger than the subject, number of marriages of each parent, patrilocal and matrilocal residence, years spent in the pastor's school and in the government school and achievement there, whether the child had ever been out of the village or off the island, sex experience, etc. The children were also given a makeshift intelligence test, colour-naming, rote memory, opposites, substitution, ball and field, and picture interpretation. These tests were all given in

Samoan; standardisation was, of course, impossible and ages were known only relatively; they were mainly useful in assisting me in placing the child within her group, and have no value for comparative purposes. The results of the tests did indicate, however, a very low variability within the group. The tests were supplemented by a questionnaire which was not administered formally but filled in by random questioning from time to time. This questionnaire gave a measure of their industrial knowledge, the extent to which they participated in the lore of the community, of the degree to which they had absorbed European teaching in matters like telling time, reading the calendar, and also of the extent to which they had participated in or witnessed scenes of death, birth, miscarriage, etc.

But this quantitative data represents the barest skeleton of the material which was gathered through months of observation of the individuals and of groups, alone, in their households, and at play. From these observations, the bulk of the conclusions are drawn concerning the attitudes of the children towards their families and towards each other, their religious interests or the lack of them, and the details of their sex lives. This information cannot be reduced to tables or to statistical statements. Naurally in many cases it was not as full as in others. In some cases it was necessary to pursue a more extensive enquiry in order to understand some baffling aspect of the child's behaviour. In all cases the investigation was pursued until I felt that I understood the girl's motivation and the degree to which her family group and affiliation in her age group explained her attitudes.

The existence of the pastor's boarding-school for girls past puberty provided me with a rough control group. These girls were so severely watched that heterosexual activities were impossible; they were grouped together with other girls of the same age regardless of relationship; they lived a more ordered and regular life than the girls who remained in their households. The ways in which they differed from other girls of the same age and more resembled European girls of the same age follow with surprising accuracy the lines suggested by the specific differences in environment. However, as they lived part of the time at home, the environmental break was not complete and their value as a control group is strictly limited.

APPENDIX III

SAMOAN CIVILISATION AS IT IS TO-DAY

The scene of this study was the little island of Taū. Along one coast of the island, which rises precipitately to a mountain peak in the centre, cluster three little villages, Lumā and Siufaga, side by side, and Faleasao, half a mile away. On the other end of the island is the isolated village of Fitiuta, separated from the other three villages by a long and arduous trail. Many of the people from the other villages have never been to Fitiuta, eight miles away. Twelve miles across the open sea are the two islands of Ofu and Olesega, which with Taū make up the Manu'a Archipelago, the most primitive part of Samoa. Journeys in slender outrigger canoes from one of these three little islands to another are frequent, and the inhabitants of Manu'a think of themselves as a unit as over against the inhabitants of Tutuila, the large island where the Naval Station is situated. The three islands have a population of a little over two thousand people, with constant visiting, inter-marrying, adoption going on between the seven villages of the Archipelago.

The natives still live in their beehive-shaped houses with floors of coral rubble, no walls except perishable woven blinds which are lowered in bad weather, and a roof of sugar-cane thatch over which it is necessary to bind palm branches in every storm. They have substituted cotton cloth for their laboriously manufactured bark cloth for use as everyday clothing, native costume being reserved for ceremonial occasions. But the men content themselves with a wide cotton loin cloth, the *lavalava*, fastened at the waist with a dexterous twist of the material. This costume permits a little of the tattooing which covers their bodies from knee to the small of the back, to appear above and below the folds of the *lavalava*. Tattooing has been taboo on Manu'a for two generations, so only a part of the population have made the necessary journey to another island in search of a tattooer. Women wear a longer *lavalava* and a short cotton dress falling to their knees. Both sexes go barefoot and hats are worn only to Church, on which occasions the men don white shirts and white coats, ingeniously tailored by the native women in imitation of Palm Beach coats which have fallen into their hands. The women's tattooing is much sparser than the men's, a mere matter of dots and crosses on arms, hands, and thighs. Garlands of flowers, flowers in the hair, and flowers twisted about the ankles, serve to relieve the drabness of the faded cotton clothing, and on gala days, beautifully patterned bark cloth, fine mats, gaily bordered with red parrot feathers, headdresses of human hair decorated with plumes and feathers, recall the more picturesque attire of pre-Christian days.

Sewing-machines have been in use for many years, although the natives are still dependent upon some deft-handed sailor for repairs. Scissors have also been added to the household equipment, but wherever possible a Samoan woman still uses her teeth or a piece of bamboo. At the Missionary boarding-schools a few of the women have learned to crochet and embroider, using their skill particularly to ornament the plump, hard pillows which are rapidly displacing the little bamboo head rests. Sheets of white cotton have taken the place of sheets of firmly woven mats or of bark cloth. Mosquito nets of cotton netting make a native house much more endurable than must have been the case when bark cloth tents were the only defence against insects. The netting is suspended at night from stout cords hung across the house, and the edges weighted down with stones, so that prowling dogs, pigs, and chickens wander through the house at will without disturbing the sleepers.

Agate buckets share with hollowed cocoanut shells the work of bringing water from the springs and from the sea, and a few china cups and glasses co-operate with the cocoanut drinking cups. Many households have an iron cook pot in which they can boil liquids in preference to the older method of dropping red hot stones into a wooden vessel containing the liquid to be heated. Kerosene lamps and lanterns are used extensively; the old candle-nut clusters and cocoanut oil lamps being reinstated only in times of great scarcity when they cannot afford to purchase kerosene. Tobacco is a much-prized luxury; the Samoans have learned to grow it, but imported varieties are very much preferred to their own.

Outside the household the changes wrought by the introduction of European articles are very slight. The native uses an iron knife to cut his copra and an iron adze blade in place of the old stone one. But he still binds the rafters of his house together with cinet and sews the parts of his fishing canoes together. The building of large canoes has been abandoned. Only small canoes for fishing are built now, and for hauling supplies over the reef the natives build keeled rowboats. Only short voyages are made in small canoes and rowboats, and the natives wait for the coming of the Naval ship to do their traveling. The government buys the copra and with the money so obtained the Samoans buy cloth, thread, kerosene, soap, matches, knives, belts, and tobacco, pay their taxes levied on every man over a certain height (as age is an indeterminate matter), and support the church.

And yet, while the Samoans use these products of a more complex civilisation, they are not dependent upon them. With the exception of making and using stone tools, it is probably safe to say that none of the native arts have been lost. The women all make bark cloth and weave fine mats. Parturition still takes place on a piece of bark cloth, the umbilical cord is cut with a piece of bamboo, and the new baby is wrapped in a specially prepared piece of white bark cloth. If soap cannot be obtained, the wild orange provides a frothy substitute. The men still manufacture their own nets, make their own hooks, weave their own eel traps. And although they use matches when they can get them, they have not lost the art of converting a carrying stick into a fire plow at a moment's notice.

Perhaps most important of all is the fact that they still depend entirely upon their own foods, planted with a sharpened pole in their own plantations. Breadfruit, bananas, taro, yams, and cocoanuts form a substantial and monotonous accompaniment for the fish, shell fish, land crabs, and occasional pigs and chickens. The food is carried down to the village in baskets, freshly woven from palm leaves. The cocoanuts are grated on the end of a wooden "horse," pointed with shell or iron; the breadfruit and taro are supported on a short stake, tufted with cocoanut husk, and the rind is grated off with a piece of cocoanut shell. The green bananas are skinned with a bamboo knife. The whole amount of food for a family of fifteen or twenty for two or three days is cooked at once in a large circular pit of stones. These are first heated to white heat; the ashes are then raked away; the food placed on the stones and the oven covered with green leaves, under which the food is baked thoroughly. Cooking over, the food is stored in baskets which are hung up inside the main house. It is served on palm leaf platters, garnished with a fresh banana leaf.

Fingers are the only knives and forks, and a wooden finger bowl is passed ceremoniously about at the end of the meal.

Furniture, with the exception of a few chests and cupboards, has not invaded the house. All life goes on on the floor. Speaking on one's feet within the house is still an unforgivable breach of etiquette, and the visitor must learn to sit cross-legged for hours without murmuring.

The Samoans have been Christian for almost a hundred years. With the exception of a small number of Catholics and Mormons, all the natives of American Samoa are adherents of the London Missionary Society, known in Samoa as the "Church of Tahiti," from its local origin. The Congregationalist missionaries have been exceedingly successful in adapting the stern doctrine and sterner ethics of a British Protestant sect to the widely divergent attitudes of a group of South Sea islanders. In the Missionary boarding-schools they have trained many boys as native pastors and as missionaries for other islands, and many girls to be the pastors' wives. The pastor's house is the educational as well as the religious center of the village. In the pastor's school the children learn to read and write their own language, to which the early missionaries adapted our script, to do simple sums and sing hymns. The missionaries have been opposed to teaching the natives English, or in any way weaning them away from such of the simplicity of their primitive existence as they have not accounted harmful. Accordingly, although the elders of the church preach excellent sermons and in many cases have an extensive knowledge of the Bible (which has been translated into Samoan), although they keep accounts, and transact lengthy business affairs, they speak no English, or only very little of it. On Taū there were never more than half a dozen individuals at one time who had any knowledge of English.

The Naval Government has adopted the most admirable policy of benevolent non-interference in native affairs. It establishes dispensaries and conducts a hospital where native nurses are trained. These nurses are sent out into the villages where they have surprising success in the administration of the very simple remedies at their command, castor oil, iodine, argyrol, alcohol rubs, etc. Through periodic administrations of salvarsan the more conspicuous symptoms of yaws are rapidly disappearing. And the natives are learning to come to the dispensaries for medicine rather than aggravate conjunctivitis to blindness by applying irritating leaf poultices to the inflamed eyes.

Reservoirs have been constructed in most of the villages, providing an unpolluted water supply at a central fountain where all the washing and bathing is done. Copra sheds in each village store the copra until the government ship comes to fetch it. Work on copra sheds, on village boats used in hauling the copra over the reef, on roads between villages, on the repairs of the water system, is carried through by a levy upon the village as a whole, conforming perfectly to the native pattern of communal work. The government operates through appointed district governors and county chiefs, and elected "mayors" in each village. The administrations of these officials are peaceful and effective in proportion to the importance of their rank in the native social organisation. Each village also has two policemen who act as town criers, couriers on government inspections, and carriers of the nurses' equipment from village to village. There are also county judges. A main court is presided over by an American civil judge and a native judge. The penal code is a random combination of government edicts, remarkable for their tolerance of native custom. When no pronouncement on a point of law is found in this code, the laws of the state of California, literally interpreted and revised, are used to provide a legalistic basis for the court's decision. These courts have taken over the settlement of disputes concerning important titles, and property rights, and the chief causes of litigation in the "courthouse" at Pago Pago are the same which agitated the native *fonos* some hundred years ago.

Schools are now maintained in many villages, where the children, seated cross-legged on the floor of a large native house, learn the haziest of English from boys whose knowledge of the language is little more extensive than theirs. They also learn part singing, at which they are extraordinarily adept, and to play cricket and many other games. The schools are useful in instilling elementary ideas of hygiene, and in breaking down the barriers between age and sex groups and narrow residential units. From the pupils in the outlying schools the most promising are selected to become nurses, teachers, and candidates for the native marine corps, the *Fitafitas,* who constitute the police, hospital corpsmen and interpreters for the naval administration. The Samoans' keen feeling for social distinction makes them particularly able to co-operate with a government in which there is a hierarchy of officialdom; the shoulder stars and bars are fitted into their own system of rank without confusion. When the Governor and group of officers pay an official visit, the native talking chief distributes the kava, first to the Governor, then to the highest chief among the hosts, then to the Commander of the Naval Yard, then to the next highest chief, without any difficulty.

In all the descriptions of Samoan life, one of the points which must have struck the reader most forcibly is the extreme flexibility of the civilisation as it is found to-day. This flexibility is the result of the blending of the various European ideas, beliefs, mechanical devices, with the old primitive culture. It is impossible to say whether it is due to some genius in the Samoan culture itself, or to fortunate accident, that these foreign elements have received such a thorough and harmonious acculturation. In many parts of the South Seas contact with white civilisation has resulted in the complete degeneration of native life, the loss of native techniques, and traditions, and the annihilation of the past. In Samoa this is not so. The growing child is faced by a smaller dilemma than that which confronts the American-born child of European parentage. The gap between parents and children is narrow and painless, showing few of the unfortunate aspects usually present in a period of transition. The new culture, by offering alternative careers to the children has somewhat lightened the parental yoke. But essentially the children are still growing up in a homogeneous community with a uniform set of ideals and aspirations. The present ease of adolescence among Samoan girls which has been described cannot safely be attributed to a period of transition. The fact that adolescence can be a period of unstressed development is just as significant. Given no additional outside stimulus or attempt to modify conditions, Samoan culture might remain very much the same for two hundred years.

But it is only fair to point out that Samoan culture, before white influence, was less flexible and dealt less kindly with the individual aberrant. Aboriginal Samoa was harder on the girl sex delinquent than is present-day Samoa. And the reader must not mistake the conditions which have been described for the aboriginal ones, nor for typical primitive ones. Present-day Samoan civilisation is simply the result of the fortuitous and on the whole fortunate impetus of a complex, intrusive culture upon a simpler and most hospitable indigenous one.

In former times, the head of the household had life and death powers over every individual under his roof. The American legal system and the missionary teachings between them have outlawed and banished these rights. The individual still benefits by the communal ownership of property, by the claims which he has on all family land; but he no longer suffers from an irksome tyranny which could be enforced with violence and possible death. Deviations from chastity were formerly punished in the case of girls by a very severe beating and a stigmatising shaving of the head. Missionaries have discouraged the beating and head shaving, but failed to substitute as forceful an inducement to circumspect conduct. The girl whose activities are frowned upon by her family is in a far better position than that of her great-grandmother. The navy has prohibited, the church has interdicted the defloration ceremony, formerly an inseparable part of the marriages of girls of rank; and thus the most potent inducement to virginity has been abolished. If for these cruel and primitive methods of enforcing a stricter régime there had been substituted a religious system which seriously branded the sex offender, or a legal system which prosecuted and punished her, then the new hybrid civilisation might have been as heavily fraught with possibilities of conflict as the old civilisation undoubtedly was.

This holds true also for the ease with which young people change their residence. Formerly it might have been necessary to flee to a great distance to avoid being beaten to death. Now the severe beatings are deprecated, but the running-away pattern continues. The old system of succession must have produced many heartburns in the sons who did not obtain the best titles; to-day two new professions are open to the ambitious, the ministry and the *Fitafitas*. The taboo system, although never as rigorous in Samoa as in other parts of Polynesia, undoubtedly compelled the people to lead more circumspect lives and stressed more vividly difference in rank. The few economic changes which have been introduced have been just sufficient to slightly upset the system of prestige which was based on display and lavish distribution of property. Acquiring wealth is easier, through raising copra, government employment, or manufacturing curios for the steamer-tourist trade on the main island. Many high chiefs do not find it worth while to keep up the state to which they are entitled, while numerous upstarts have an opportunity to acquire prestige denied to them under a slower method of accumulating wealth. The intensity of local feeling with its resulting feuds, wars, jealousies and conflicts (in the case of intermarriages between villages) is breaking down with the improved facilities for transportation and the co-operation between villages in religious and educational matters.

Superior tools have partially done away with the tyranny of the master craftsman. The man who is poor, but ambitious, finds it easier to acquire a guest house than it would have been when the laborious highly specialised work was done with stone tools. The use of some money and of cloth, purchased from traders, has freed women from part of the immense labour of manufacturing mats and tapa as units of exchange and for clothing. On the other hand, the introduction of schools has taken an army of useful little labourers out of the home, especially in the case of the little girls who cared for the babies, and so tied the adult women more closely to routine domestic tasks.

Puberty was formerly much more stressed than it is to-day. The menstrual taboos against participation in the kava ceremony and in certain kinds of cooking were felt and enforced. The girl's entrance into the *Aualuma* was always, not just occasionally, marked by a feast. The unmarried girls and the widows slept, at least part of the time, in the house of the *taupo*. The *taupo* herself had a much harder life. To-day she pounds the kava root, but in her mother's day it was chewed until jaws ached from the endless task. Formerly, should a defection from chastity be disclosed at her marriage, she faced being beaten to death. The adolescent boy faced tattooing, a painful, wearisome proceeding, additionally stressed by group ceremony and taboo. To-day, scarcely half of the young men are tattooed; the tattooing is performed at a much more advanced age and has no connection with puberty; the ceremonies have vanished and it has become a mere matter of a fee to the artist.

The prohibitions against blood revenge and personal violence have worked like a yeast in giving greater personal freedom. As many of the crimes which were formerly punished in this fashion are not recognised as crimes by the new authorities, no new mechanism of punishment has been devised for the man who marries the divorced wife of a man of higher rank, the miscreant who gossips outside his village and so brings his village into disrepute, the insolent detractor who recites another's genealogy, or the naughty boy who removes the straws from the pierced cocoanuts and thus offers an unspeakable affront to visitors. And the Samoan is not in the habit of committing many of the crimes listed in our legal code. He steals and is fined by the government as he was formerly fined by the village. But he comes into very slight conflict with the central authorities. He is too accustomed to taboos to mind a quarantine prohibition which parades under the same guise; too accustomed to the exactions of his relations to fret under the small taxation demands of the government. Even the stern attitude formerly taken by the adults towards precocity has now been subdued, for what is a sin at home becomes a virtue at school.

The new influences have drawn the teeth of the old culture. Cannibalism, war, blood revenge, the life and death power of the *matai,* the punishment of a man who broke a village edict by burning his house, cutting down his trees, killing his pigs, and banishing his family, the cruel defloration ceremony, the custom of laying waste plantations on the way to a funeral, the enormous loss of life in making long voyages in small canoes, the discomfort due to wide-spread disease—all these have vanished. And as yet their counterparts in producing misery have not appeared.

Economic instability, poverty, the wage system, the separation of the worker from his land and from his tools, modern warfare, industrial disease, the abolition of leisure, the irksomeness of a bureaucratic government—these have not yet invaded an island without resources worth exploiting. Nor have the subtler penalties of civilisation, neuroses, philosophical perplexities, the individual tragedies due to an increased consciousness of personality and to a greater specialisation of sex feeling, or conflicts between religion and other ideals, reached the natives. The Samoans have only taken such parts of our culture as made their life more comfortable, their culture more flexible, the concept of the mercy of God without the doctrine of original sin.

APPENDIX IV

THE MENTALLY DEFECTIVE AND THE MENTALLY DISEASED

Without any training in the diagnosis of the mentally diseased and without any apparatus for exact diagnosis of the mentally defective, I can simply record a number of amateur observations which may be of interest to the specialist interested in the possibilities of study-ing the pathology of primitive peoples. In the Manu'a Archipelago with a population of a little over two thousand people, I saw one case which would be classed as idiocy, one imbecile, one boy of fourteen who appeared to be both feeble-minded and insane, one man of thirty who showed a well-systematised delusion of grandeur, and one sexual invert who approximated in a greater development of the breasts, mannerism and attitudes of women and a preference for women's activities, to the norm of the opposite sex. The idiot child was one of seven children; he had a younger brother who had walked for over a year, and the mother declared that there were two years between the children. His legs were shrunken and withered, he had an enormous belly and a large head set very low on his shoulders. He could neither walk nor talk, drooled continually, and had no command over his excretory functions. The imbecile girl lived on another island and I had no opportunity to observe her over any length of time. She was one or two years past puberty and was pregnant at the time that I saw her. She could talk and perform the simple tasks usually performed by children of five or six. She seemed to only half realise her condition and giggled foolishly or stared vacantly when it was mentioned. The fourteen-year-old boy was at the time when I saw him definitely demented, giving an external picture of catatonic dementia præcox. He took those attitudes which were urged upon him, at times, however, becoming violent and unmanageable. The relatives insisted that he had always been stupid but only recently become demented. For this I have only their word as I was only able to observe the boy during a few days. In no one of these three cases of definite mental deficiency was there any family history which threw any light upon the matter. Among the girls whom I studied in detail only one, Sala, discussed in Chapter X, was sufficiently inferior to the general norm of intelligence to approximate to a moron.

The man with the systematised delusion of grandeur was said to be about thirty years of age. Gaunt and emaciated, he looked much older. He believed that he was Tufele, the high chief of another island and the governor of the entire archipelago. The natives conspired against him to rob him of his rank and to exalt an usurper in his stead. He was a member of the Tufele family but only very remotely so that his delusion bore no relation to reality as he would never have had any hope of succeeding to the title. The natives, he said, refused to give him food, mocked him, disallowed his claims, did their best to destroy him, while a few white people were wise enough to recognise his rank. (The natives instructed visitors to address him in the chief's language because he consented to dance, a weird pathetic version of the usual style, only when so opportuned.) He had no outbreaks of violence, was morose, recessive, only able to work at times and never able to do heavy work or to be trusted to carry through any complicated task. He was treated with universal gentleness and toleration by his relatives and neighbours.

From informants I obtained accounts of four cases on Tutuila which sounded like the manic stage of manic depressive insanity. All four of these individuals had been violently destructive, and uncontrollable for a period of time, but had later resumed what the natives considered normal functioning. An old woman who had died some ten years before was said to have compulsively complied with any command that was given her. There was one epileptic boy in Taū, a member of an otherwise normal family of eight children. He fell from a tree during a seizure and died from a fractured skull soon after I came to Manu'a. A little girl of about ten who was paralysed from the waist down was said to be suffering from an overdose of salvarsan and to have been normal until she was five or six years old.

Only two individuals, one a married woman of thirty or so, the other a girl of nineteen, discussed in Chapter X, showed a definite neurasthenic constitution. The married woman was barren and spent a great deal of time explaining her barrenness as need of an operation. The presence of an excellent surgeon at the Samoan hospital during the preceding two years had greatly enhanced the prestige of operations. On Tutuila, near the Naval Station, I encountered several middle-aged women obsessed with operations which they had undergone or were soon to undergo. Whether this vogue of modern surgery, by giving it special point, has added to the amount of apparent neurasthenia or not, it is impossible to say.

Of hysterical manifestations, I encountered only one, a girl of fourteen or fifteen with a bad tic in the right side of her face. I only saw her for a few minutes on a journey and was unable to make any investigations. I neither saw nor heard of any cases of hysterical blindness or deafness, nor of any anæsthesias nor paralyses.

I saw no cases of cretinism. There were a few children who had been blind from birth. Blindness, due to the extremely violent methods used by the native practitioners in the treatment of "Samoan conjunctivitis," is common.

The pathology which is immediately apparent to any visitor in a Samoan village is mainly due to the diseased eyes, elephantiasis, and abscesses and sores of various sorts, but the stigmata of degeneration are almost entirely absent.

There was one albino, a girl of ten, with no albinism in the recorded family history, but as one parent, now dead, had come from another island, this was not at all conclusive data.

APPENDIX V

MATERIALS UPON WHICH THE ANALYSIS IS BASED

This study included sixty-eight girls between the ages of eight and nine and nineteen or twenty—all the girls between these ages in the three villages of Faleasao, Lumā and Siufaga, the three villages on the west coast of the island of Taū in the Manu'a Archipelago of the Samoan Islands.

Owing to the impossibility of obtaining accurate dates of birth except in a very few cases, the ages must all be regarded as approximate. The approximations were based upon the few known ages and the testimony of relatives as to the relative age of the others. For purpose of description and analysis I have divided them roughly into three groups, the children who showed no mammary signs of puberty, twenty-eight in number, ranging in age from eight or nine to about twelve or thirteen; the children who would probably mature within the next year or year and a half, fourteen in number, ranging in age from twelve or thirteen to fourteen or fifteen; and the girls who were past puberty, but who were not yet considered as adults by the community, twenty-five in number, ranging in age from fourteen or fifteen to nineteen or twenty. These two latter groups and eleven of the younger children were studied in detail, making a group of fifty. The remaining fourteen children in the youngest group were studied less carefully as individuals. They formed a large check group in studying play, gang life, the development of brother and sister avoidance, the attitude between the sexes, the difference in the interests and activities of this age and the girls approaching puberty. They also provided abundant material for the study of the education and discipline of the child in the home. The two tables present in summary form the major statistical facts which were gathered about the children specially studied, order of birth, number of brothers and sisters, death or remarriage or divorce of parents, residence of the child, type of household in which the child lived and whether the girl was the daughter of the head of the household or not. The second table relates only to the twenty-five girls past puberty and gives length of time since first menstruation, frequency of menstruation, amount and locations of menstrual pain, the presence or absence of masturbation, homosexual and heterosexual experience, and the very pertinent fact of residence or non-residence in the pastor's household. A survey of the summary analyses joined to these tables will show that these fifty girls present a fairly wide range in family organisation, order of birth, and relation to parents. The group may be fairly considered as representative of the various types of environment, personal and social, which are found in Samoan civilisation as it is to-day.

DISTRIBUTION OF GROUPS OF ADOLESCENTS IN RELATION TO
FIRST MENSTRUATION

Within last six months	6
Within last year	3
Within last two years	5
Within last three years	7
Within last four years	3
Within last five years	1
Total	25

Household number   Girl's number   Name   Age   (How estimated)
Matai   Rank   Father   Rank   Father's residence
Mother   Residence of mother   Either parent been married before?
Economic status of household   Church membership of father, mother, guardian
Menstruated?   Date of commencement?   Pain   Regularity   Estimate of physical development
Grade in government school?   In pastor's school?   Any knowledge of English?
Foreign experience (outside Tau)   Physical defects
Order of birth?
Best friends in order?

Test Scores   Religious attitudes
  Colour naming
  Rote memory for digits
  Digit symbol substitution
  Opposites
  Picture Interpretation
  Ball and Field

Judgments on individuals in the village   Personality
  Most beautiful girl
  Handsomest boy
  Wisest man
  Cleverest woman   Attitude towards household
  Worst boy
  Worst girl
  Best boy
  Best girl   Attitude towards contemporaries

TABLE I

TABLE SHOWING LENGTH OF TIME SINCE PUBERTY, PERIODICITY, AMOUNT OF PAIN DURING MENSES, MASTURBATION, HOMOSEXUAL EXPERIENCE, HETEROSEXUAL EXPERIENCE, AND RESIDENCE OR NON-RESIDENCE IN PASTOR'S HOUSEHOLD

No.	Name	Time Elapsed Since Puberty	Periodicity	Pain*	Masturbation	Homosexual Experience	Heterosexual Experience	Residence in Pastor's Household
1.	Luna	3 years	monthly	abdo.	yes	yes	yes	no
2.	Masina	3 "	"	"	"	"	"	"
3.	Losa	2 "	"	abdo. back	no	"	no	yes
4.	Sona	3 "	semi-monthly	" "	yes	"	"	"
5.	Loto	2 months	monthly	back	"	"	"	"
6.	Pala	6 "	"	none	"	"	"	no
7.	Aso	18 "	semi-monthly	back	"	no	"	"
8.	Tolo	3 "	" "	extreme	"	"	"	"
9.	Lotu	3 years	monthly	"	"	yes	yes	"
10.	Tulipa	2 months	"	abdo. back	"	"	no	yes
14.	Lita	2 years	"	back	"	"	"	no
16.	Namu	3 "	"	"	"	"	yes	"
17.	Ana	2 "	Every three months	"	"	"	no	yes
18.	Lua	3 months	monthly	"	no	no	"	no
19.	Tolu	4 years	semi-monthly	"	yes	yes	yes	"
21.	Mala	2 months	monthly	"	"	no	no	"
22.	Fala	1 year	"	"	"	yes	yes	"
23.	Lola	1 "	semi-monthly	abdo.	"	"	"	"
23a.	Tulipa	3 years	monthly	back	"	"	"	"

* Abdomen—pain only there; back—pain only there; extreme—so characterised by girl, never so ill that she couldn't work.

No.	Name							
24.	Leta	2 months	monthly	none	yes	yes	yes	yes
25.	Ela	2 years	"	extreme	"	"	"	"
27.	Mina	5 "	"	"	"	no	no	"
28.	Moana	4 "	bi-monthly	abdo. back	"	"	yes	no
29.	Luina	4 "	monthly	extreme	no	"	no	yes
30.	Sala	3 "	semi-monthly	"	yes	"	yes	no

TABLE II

FAMILY STRUCTURE (See key to this table on p. 171)

No.	Name	1	2	3	4	5	6	7	8	9	10	11	12	13	14	15	16	17	18	19	20	21
Pre-Ads.																						
1.	Tuna	1	3																		x	x
2.	Vala					1–	3–		x			x			x						x	
3.	Pele	3	4																		x	
4.	Timu							x	x											x		
5.	Suna							x				x								x		
6.	Pola		3	2	1										x						x	
7.	Tua	1	4	1												x						
8.	Sina	1	1	2	3																x	x
9.	Fiva	1		1	3														x			
10.	Ula	1	1	1	2															x		x
11.	Siva	1	4						x										x			
Midways																						
1.	Tasi	1		4									x								x	x
2.	Fitu	1		2	2										x							x
3.	Mata	1	1		3								x								x	x

4.	Vi	3	3	1	1									x			
6.	Ipu	2	1				x		x						x		
7.	Selu	3															
8.	Pula	2		1	1			x							x		x
9.	Meta	3		1	1										x		
10.	Maliu		2	2	2				x							x	
11.	Fiatia			3–	2–			x	x		½	½	x				
12.	Lama	3				x							x				
13.	Tino	1		2	1			x				x	x				
14.	Vina	1	2	2	1											x	x
15.	Talo			2	4									x			
Adolescents																	
1.	Luna	2	5		1				x		x		x		x		
2.	Masina	3		2	2		x						x				
3.	Losa			2	1				x						x	x	
4.	Sona	2				x		x					x				
5.	Loto	4	1			x	x						x				
6.	Pala	3	3	1											x		
7.	Iso	1	3	1			x				x						
8.	Tolo	1	2				x						x				
9.	Lotu			3	5											x	x
10.	Tulipa	5	3										x				
14.	Lita	4		2	1								x				
16.	Namu			4	2				x				x		x	x	
17.	Ana			3–				x					x				

TABLE II
FAMILY STRUCTURE (*Concl.*)

No.	Name	1	2	3	4	5	6	7	8	9	10	11	12	13	14	15	16	17	18	19	20	21
18.	Lua			7	1												*x*		*x*			
19.	Tolu								*x*										*x*			
21.	Mala	3	1					*x*	*x*			*x*	*x*			*x*						
22.	Fala	1	3	3	1				*x*							*x*			*x*			
23.	Lola	1	2		2			*x*					*x*					*x*		*x*		
23a.	Tulipa	2	2											*x*								
24.	Leta	1	4						*x*							*x*						
25.	Ela	2	1	1				*x*				*x*		*x*		*x*						
27.	Mina		1					*x*			*x*					*x*					*x*	*x*
28.	Moana	1	4	1		1x	1x				*x*		*x*	*x*								
29.	Luina						1	*x*								*x*						
30.	Sala	3	1					*x*														

Column	Subject

1 Number of older brothers
2 Number of older sisters
3 Number of younger brothers
4 Number of younger sisters
5 Half brother, *plus*, number older, *minus*, number younger
6 Half sister, *plus*, number older, *minus*, number younger
7 Mother dead
8 Father dead
9 Child of mother's second marriage
10 Child of father's second marriage
11 Mother remarried
12 Father remarried
13 Residence with both parents and patrilocal
14 Residence with both parents and matrilocal
15 Residence with mother only
16 Residence with father only
17 Parents divorced
18 Residence with paternal relatives
19 Residence with maternal relatives
20 Father is *matai* of household
21 Residence in a biological family, i.e., household of parents, children, and no more than two additional relatives.

x in the table means the presence of trait. For example, *x* in column 7 means that the mother is dead.

APPENDIX V

ANALYSIS OF TABLE ON FAMILY STRUCTURE

There were among the sixty-eight girls:

7 only children
15 youngest children
5 oldest children
5 with half brother or sister in the same household
5 whose mother was dead
14 whose father was dead
3 who were children of mother's second marriage
2 children of father's second marriage
7 whose mother had remarried
5 whose father had remarried
4 residence with both parents patrilocal
8 residence with both parents matrilocal
9 residence with mother only
1 residence with father only
7 parents divorced
12 residence with paternal relatives (without either parent)
6 residence with maternal relatives (without either parent)

15, or 30% whose fathers were heads of households

12 who belonged to a qualified biological family (i.e., a family which during my stay on the island comprised only two relatives beside the parents and children).

INTELLIGENCE TESTS USED

It was impossible to standardise any intelligence tests and consequently my results are quantitatively valueless. But as I had had some experience in the diagnostic use of tests, I found them useful in forming a preliminary estimate of the girls' intelligence. Also, the natives have long been accustomed to examinations which the missionary authorities conduct each year, and the knowledge that an examination is in progress makes them respect the privacy of investigator and subject. In this way it was possible for me to get the children alone, without antagonizing their parents. Furthermore, the novelty of the tests, especially the colour-naming and picture interpretation tests, served to divert their attention from other questions which I wished to ask them. The results of the tests showed a much narrower range than would be expected in a group varying in age from ten to twenty. Without any standardisation it is impossible to draw any more detailed conclusions. I shall, however, include a few comments about the peculiar responses which the girls made to particular tests, as I believe such comment is useful in evaluating intelligence testing among primitive peoples and also in estimating the possibilities of such testing.

Tests Used

Colour Naming. 100 half-inch squares, red, yellow, black and blue.

Rote Memory for Digits. Customary Stanford Binet directions were used.

Digit Symbol Substitution. 72 one-inch figures, square, circle, cross, triangle and diamond.

Opposites. 23 words. Stimulus words: fat, white, long, old, tall, wise, beautiful, late, night, near, hot, win, thick, sweet, tired, slow, rich, happy, darkness, up, inland, inside, sick.

Picture Interpretation. Three reproductions from the moving picture *Moana*, showing, (*a*) Two children who had caught a cocoanut crab by smoking it out of the rocks above them, (*b*) A canoe putting out to sea after bonito as evidenced by the shape of the canoe and the position of the crew, (*c*) A Samoan girl sitting on a log eating a small live fish which a boy, garlanded and stretched on the ground at her feet, had given her.

Ball and Field. Standard-sized circle.

Standard directions were given throughout in all cases entirely in Samoan. Many children, unused to such definitely set tasks, although all are accustomed to the use of slate and of pencil and paper, had to be encouraged to start. The ball and field test was the least satisfactory as in over fifty per cent of the cases the children followed an accidental first line and simply completed an elaborate pattern within the circle. When this pattern happened by accident to be either the Inferior or Superior solution, the child's comment usually betrayed the guiding idea as æsthetic rather than as an attempt to solve the problem. The children whom I was led to believe to be most intelligent, subordinated the æsthetic consideration to the solution of the problem, but the less intelligent children were sidetracked by their interest in the design they could make much more easily than are children in our civilisation. In only two cases did I find a rote memory for digits which exceeded six digits; two girls completing seven successfully. The Samoan civilisation puts the slightest of premiums upon rote memory of any sort. On the digit-symbol test they were slow to understand the point of the test and very few learned the combinations before the last line of the test sheet. The picture interpretation test was the most subject to vitiation through a cultural factor; almost all of the children adopted some highly stylised form of comment and then pursued it through one balanced sentence after another: "Beautiful is the boy and beautiful is the girl. Beautiful is the garland of the boy and beautiful is the wreath of the girl," etc. In the two pictures which emphasised human beings no discussion could be commenced until the question of the relationship of the characters had been ascertained. The opposites test was the one which they did most easily, a natural consequence of a vivid interest in words, an interest which leads them to spend most of their mythological speculation upon punning explanations of names.

CHECK LIST USED IN INVESTIGATION OF EACH
GIRL'S EXPERIENCE

In order to standardise this investigation I made out a questionnaire which I filled out for each girl. The questions were not asked consecutively but from time to time I added one item of information after another to the record sheets. The various items fell into the loose groupings indicated below.

Agricultural proficiency. Weeding, selecting leaves for use in cooking, gathering bananas, taro, breadfruit, cutting cocoanuts for copra.

Cooking. Skinning bananas, grating cocoanut, preparing breadfruit, mixing *palusami*,* wrapping *palusami*, making *tafolo*,† making banana *poi*, making arrow-root pudding.

Fishing. Daylight reef fishing, torchlight reef fishing, gathering *lole*, catching small fish on reef, using the "come hither" octopus stick, gathering large crabs.

Weaving. Balls, pin-wheels, baskets to hold food gifts, carrying baskets, woven blinds, floor mats, fishing baskets, food trays, thatching mats, roof bonnetting mats, plain fans, pandanus

---

* *Palusami*—a pudding prepared from grated cocoanut, flavored with red hot stone, mixed with sea water, and wrapped in taro leaves, from which the acrid stem has been scorched, then in a banana leaf, finally in a breadfruit leaf.
† *Tafolo*—a pudding made of breadfruit with a sauce of grated cocoanut.

173

floor mats, bed mats (number of designs known and number of mats completed), fine mats, dancing skirts, sugar-cane thatch.

Bark cloth making. Gathering paper mulberry wands, scraping the bark, pounding the bark, using a pattern board, tracing patterns free hand.

Care of clothing. Washing, ironing, ironing starched clothes, sewing, sewing on a machine, embroidering.

Athletics. Climbing palm trees, swimming, swimming in the swimming hole within the reef,* playing cricket.

Kava making. Pounding the kava root, distributing the kava, making the kava, shaking out the hibiscus bark strainer.

Proficiency in foreign things. Writing a letter, telling time, reading a calendar, filling a fountain pen.

Dancing.

Reciting the family genealogy.

Index of knowledge of the courtesy language. Giving the chiefs' words for: arm, leg, food, house, dance, wife, sickness, talk, sit. Giving courtesy phrases of welcome, when passing in front of some one.

Experience of life and death. Witnessing of birth, miscarriage, intercourse, death, Cæsarian post-mortem operation.

Marital preferences, rank, residence, age of marriage, number of children.

Index of knowledge of the social organisation. Reason for Cæsarian post-mortem, proper treatment of a chief's bed, exactions of the brother and sister taboo, penalties attached to cocoanut *tapui,*† proper treatment of a kava bowl, the titles and present incumbents of the titles of the *Manaia* of Lumā, Siufaga and Faleasao, the Taupo of Fitiuta, the meaning of the *Fale Ula,*** the *Umu Sa,*†† the *Mua o le taule'ale'a*‡ the proper kinds of property for a marriage exchange, who was the high chief of Lumā, Siufaga, Faleasao and Fitiuta, and what constituted the Lafo§ of the talking chief.

---

* Swimming in the hole within the reef required more skill than swimming in still water; it involved diving and also battling with a water level which changed several feet with each great wave.

† *Tapui.* The hieroglyphic signs used by the Samoans to protect their property from thieves. The *tapui* calls down an automatic magical penalty upon the transgressor. The penalty for stealing from property protected by the cocoanut *tapui* is boils.

** The ceremonial name of the council house of the Tui Manu'a.

†† The sacred oven of food and the ceremony accompanying its presentation and the presentation of fine mats to the carpenters who have completed a new house.

‡ The ceremonial call of the young men of the village upon a visiting maiden.

§ The ceremonial perquisite of the talking chief, usually a piece of tapa, occasionally a fine mat.

Glossary of Native Terms Used in the Text

Aumaga (*'aumāga*)—the organisation of untitled men in each Samoan village.

Aualuma—the organisation of unmarried girls past puberty, wives of untitled men and widows.

Afafine—daughter (man speaking).

Aiga—relative.

Atali'i—son (man speaking).

Avaga—elopement.

Fa'alupega—the courtesy phrases, recited in formal speeches, which embody the social pattern of each village.

Fale—house.

Faletua—"she who sits in the back of the house." The courtesy term for a chief's wife.

Fono—a meeting. Specifically the organisation of titled men of a village, district or island.

Fitafita—a member of the native marine corps.

Ifo—to lower oneself to some one whom one has offended or injured.

Ifoga—the act of doing so.

Lavalava—a loin cloth, fastened by a twist in the material at the waist.

Lole—a sort of jellyfish; applied by the natives to candy.

Malaga—a travelling party; a journey.

Manaia—the heir-apparent of the principal chief; the leader of the Aumaga; the heir of any important chief whose title carries the privilege of giving a manaia title to his heir.

Matai—the holder of a title; the head of a household.

Moetotolo—surreptitious rape.

Moni—true, real.

Musu—unwillingness, obstinacy towards any course of action.

Olomatua—old woman.

Papalagi—white men; literally, "sky bursters." Foreign.

Pua—the frangipani tree.

Soa—a companion in circumcision; an ambassador in love affairs.

Soafafine—a woman ambassador in love affairs.

Siva—to dance; a dance.

Tama—a child, a son (woman speaking).

Tama—father.

Tamafafine—a child of the distaff side of the house.

Tamatane—a child of the male line.

Tapa—bark cloth.

Taule'ale'a—a member of the Aumaga; an untitled man.

Taupo—the village ceremonial hostess; the girl whom a high chief has honoured with a title and a distribution of property.

Tausi—the courtesy term for the wife of a talking chief; literally, "to care for."

Tei—a younger sibling.

Teine—a girl.

Teinetiti—a little girl.

Tinā—mother.

Toa'ina—an old man.

Tuafafine—female sibling of a male.

Tuagane—male sibling of a female.

Tulafale—a talking chief.

Uso—sibling of the same sex.

Note on the Pronunciation of Samoan Words

The vowels are all pronounced as in Italian.

G is always pronounced like NG.

The glottal stop is indicated by a (').